椭圆轨道近距离相对导航与姿轨一体化控制

Close-range Relative Navigation and Integrated Attitude/Position Control Methods for Elliptical Orbit

张力军　刘　军　张　翔　张士峰　著

国防工业出版社
·北京·

内 容 简 介

本书以近距离空间操控任务为研究背景，针对椭圆轨道近距离相对导航与姿轨一体化控制方法开展研究，系统阐述了基于单目视觉的非合作相对位姿确定方法、基于状态估计的近距离相对导航方法、基于对偶代数的航天器姿轨一体化控制方法、空间高精度姿态机动控制技术等内容，所提出的方法均适用于目标航天器轨道为椭圆或近圆轨道情形。本书注重理论性和系统性，并采用数学推导与仿真实验相结合的思路，初步解决了基于单目视觉的航天器近距离相对导航与姿轨控制中若干关键问题，具有很强的实际应用背景。

本书的主要读者对象为从事航天器相对姿态确定和控制系统设计与仿真的科研人员，同时也可作为高等院校相关专业研究生和科研学者的参考书。

图书在版编目（CIP）数据

椭圆轨道近距离相对导航与姿轨一体化控制/张力军等著. —北京：国防工业出版社，2023. 4

ISBN 978-7-118-12846-8

Ⅰ. ①椭…　Ⅱ. ①张…　Ⅲ. ①航天器-飞行控制　Ⅳ. ①V448. 2

中国国家版本馆 CIP 数据核字（2023）第 038590 号

※

国防工业出版社出版发行

（北京市海淀区紫竹院南路 23 号　邮政编码 100048）

三河市众誉天成印务有限公司印刷

新华书店经售

*

开本 710×1000　1/16　**印张** 10　**字数** 200 千字

2023 年 4 月第 1 版第 1 次印刷　**印数** 1—2000 册　**定价** 68. 00 元

国防书店：（010）88540777　　书店传真：（010）88540776

发行业务：（010）88540717　　发行传真：（010）88540762

前　　言

随着航天技术的快速发展与空间应用的不断拓展，日趋多样复杂的航天器近距离操作任务正成为未来自主航天器的重要应用方向之一，与之对应的航天器间的相对运动制导、导航和控制方法也成为近年来研究的热点。随着空间技术的进一步发展，椭圆轨道交会对接技术日益成为空间活动必需的基础技术，尤其是为了完成对失效、故障等非合作航天器的空间在轨服务，常需要其作为技术保障。目前，国内公开出版的系统研究该领域技术的书籍甚少。

近年来，在国家自然科学基金、863 计划、中国科协青年人才托举工程项目等支持下，我们围绕航天器近距离相对导航与姿轨一体化控制技术及应用做了大量研究工作，本书是根据作者在科研与实际工作中的积累而写成的，书中大部分内容取自作者公开发表的学术论文、有关技术报告和博士学位论文，并以航天器近距离操作任务为主要研究背景，深入研究了椭圆轨道近距离相对导航与姿轨一体化控制方法，以参数估计理论、非线性滤波理论、对偶代数理论、高斯伪谱法为理论基础，对基于单目视觉的非合作相对位姿确定方法、基于状态估计的近距离相对导航方法、基于对偶代数的航天器姿轨一体化控制方法、空间高精度姿态机动控制技术等做了全面系统的论述，所提出的方法均适用于目标航天器轨道为椭圆或近圆轨道情形。

全书共 6 章。第 1 章为概论，主要介绍了航天器近距离相对导航与姿轨一体化控制发展现状及关键技术进展；第 2 章介绍了本书中用到的椭圆轨道航天器相对动力学、敏感器和执行机构模型；第 3 章提出了基于标志点静态几何关系的航天器相对位姿确定方法；第 4 章论述了基于状态估计的近距离相对导航方法；第 5 章提出了基于对偶代数的航天器姿轨一体化控制方法；第 6 章论述了空间高精度姿态机动控制方法。

本书的内容紧密结合工程技术专业，采用数学推导与仿真实验相结合的思路，保证了数学模型和算法的正确性，初步解决了基于单目视觉的航天器近距离相对导航与姿轨控制中若干关键问题，取得了一些具有创新性的研究成果，具有很强的实际应用背景。本书可作为高等院校航空航天、电子信息、控制科学等专业研究生和科研学者的教学参考书，也可供从事航空宇航科学与技术、控制科学与工程领域的工程技术人员和研究人员学习参考。

本书在编写过程中得到了中国西安卫星测控中心和国防科技大学各级领导的支持与专家的推荐，在此一一表示衷心感谢。

由于作者理论和学术水平有限，书中难免有不足或疏漏之处，恳请读者批评指正。

作　者

2022 年 5 月于西安

目　　录

第1章 绪　　论

目前，世界空间在轨服务与维护事业正处于蓬勃发展的阶段，我国空间飞行器在轨服务与维护重大项目、空间站建设与运行、先进航天技术发展与应用等对航天器在轨服务和空间非合作目标操控技术与应用研究提出了重大需求。本书对航天器近距离相对导航与姿轨一体化控制方法进行研究，所提出的方法均适用于目标航天器（主星）轨道为椭圆或近圆轨道情形。

1.1 背景和意义

当今世界，航天技术已是衡量一个国家总体科技水平和综合国力的重要标志，其在科技进步、军事应用和经济发展方面均带来显著效益，因此各国都竞相积极发展航天技术。为了保证航天器能够在复杂的空间环境中持久稳定地在轨运行，各航天大国正大力研究航天器在轨服务（On-Orbit Servicing, OOS）技术[1]，并纷纷投入大量经费开展与其相关的空间技术演示验证试验。典型计划任务如美国国家航空航天局（National Aeronautics and Space Administration，NASA）的“自主交会技术演示”（Demonstration of Autonomous Rendezvous Technology，DART）试验[2]、空军研究实验室（Air Force Research Laboratory，AFRL）的试验卫星系列项目 XSS-10[3] 和 XSS-11[4]、国防高级研究计划局（Defense Advanced Research Projects Agency，DARPA）的轨道快车（Orbital Express，OE）试验[5-6]和空间轨道清道夫（Spacecraft for the Universal Modification of Orbits/Front end Robotics Enabling Near-term Demonstration，SUMO/FREND）项目[7]、欧洲航天局（Europen Space Agency，ESA）的自动转移飞行器（Automatic Transfer of Aircraft，ATV）[8-9]、日本国家空间发展署（National Space Development Agency，NASDA）的工程试验卫星（Engineering Test Satellite-7，ETS-VII）[10-11] 和 H-2 运载器（H-2 Transfer Vehicle，HTV）[12]、瑞典与德国宇航中心（Deutsches Zentrum für Luft-und Raumfahrt，DLR）合作开展的 PRISMA 项目[13]，以及中国的天宫一号与神舟飞船交会对接[14]和空间站建立[15]等。近年来，美国、日本、俄罗斯及欧洲等都在积极探索新一代航天器在轨服务技术，并利用微型航天器和集群航天器系统开展了自主交会对接、相对位姿测量、在轨加注以及模块更换等关键技术的地面和飞行演示验证试验。

随着空间机器人抓捕、故障卫星在轨捕获与维修，以及空间碎片移除等针对非合作目标的在轨服务任务需求日益增多，传统的合作式空间操控理论已不能满足信息层面不沟通、机动行为不配合、先验信息少的非合作目标操控技术要求，各航天大国正加紧开展面向非合作航天器的在轨服务技术，并纷纷投入大量经费开展与其相关的空间技术演示验证试验，典型计划任务如表 1.1 所示。2018 年 9 月，英国萨里航天中心联合空客防务公司研制的“太空碎片清除”试验卫星，完成世界首次飞网在轨抓捕立方星技术验证，后续还将验证鱼叉捕获和拖拽帆离轨等技术。与合作目标相比，非合作目标的在轨服务技术不同主要体现在近距离逼近段，此时目标航天器上未安装用于辅助测量的合作标识且无法利用星间链路向外传输自身状态信息，从而大大增加了在轨服务技术的难度，而逼近段相对位姿测量和控制的精确与否将直接影响对接过程的成败。因此，作为自主在轨服务的关键技术，逼近段的非合作航天器相对导航技术已成为近年来的研究热点，受到国内外学者的广泛关注[16]。

表 1.1　非合作目标在轨服务计划汇总表

项目名称	国家（机构）	任务内容	测量系统	启动时间
OLEV	欧洲（ESA）	地球静止轨道航天器延寿	远场相机、双目相机	2002 年
ROGER	欧洲（ESA）	利用飞网捕获目标并离轨	双目相机、激光测距仪、变焦距相机	2003 年
DEOS	德国（DLR）	在轨服务任务	视觉相机、激光雷达	2003 年
HRV	美国（DARPA）	“哈勃”望远镜自主在轨维修	激光雷达、相机	2005 年
XSS-11	美国（AFRL）	验证空间自主交会技术	激光雷达、视频相机、星敏感器	2005 年
OMS	日本（NICT）	轨道维护	视觉相机、红外敏感器、星敏感器	2005 年
ACTS	加拿大（CSA）	航天器的自主捕获与维护	激光雷达	2006 年
SDMR	日本（JAXA）	验证利用电动力绳使非合作目标离轨的技术	GPS、双目相机、星敏感器	2006 年
FREND	美国（DARPA）	非合作目标捕获	激光雷达、立体视觉	2007 年

续表

项目名称	国家（机构）	任务内容	测量系统	启动时间
Phoenix	美国（DARPA）	在轨失效卫星抓捕	设计中	2012 年
L-Restore	美国（NASA）	验证在轨燃料加注技术	设计中	2014 年
RSGS	美国（DARPA）	验证对地球同步轨道卫星自主服务能力	设计中	2016 年

在轨服务技术经历了有人参与和无人自主两个发展阶段，现代在轨服务技术主要使用空间机器人，该技术在未来空间碎片清理、深空探测以及军事作战等领域具有广阔的应用前景。空间交会对接技术是实现在轨服务的基础，在目前已经完成的交会对接任务中，目标航天器大多都运行在圆或者近圆轨道上，而椭圆轨道交会对接技术尤其是针对失效、故障等非合作航天器的空间在轨服务技术涉及得较少。目前，美国通过 XSS-11 号试验卫星项目开展了针对椭圆轨道上非合作目标的交会、伴飞、绕飞、捕获以及对接等空间操作的演示验证试验[4]。另外，在火星采样返回任务中，Sotto 研究表明采用椭圆轨道交会方案可以使轨道器在环绕轨道捕获机动和火星逃逸机动两个关键环节中节省大量燃料，从而大大降低了任务成本[17]。

从技术难度和任务拓展性来看，椭圆轨道交会对接是空间交会对接技术发展的趋势，且传统的圆参考轨道交会对接可看作其特例形式。目前，国内外学者把空间交会对接的整个飞行过程主要划分为 4 个阶段[18]：①地面导引阶段；②自动寻的阶段；③最终逼近阶段；④对接阶段。后两个阶段一般统称为航天器近距离操作段，其中涉及的轨道和姿态的六自由度导航控制问题是当前国内外研究的一个热点和难点问题。当两航天器距离较近时，轨道和姿态控制易受误差因素影响，且为了满足对接精度的要求还需要考虑轨道和姿态之间的耦合控制问题。因此，控制系统呈现多变量耦合和强非线性的特征，这给近距离操作阶段的轨道和姿态控制带来很大挑战。

1.2　航天器相对导航与姿轨耦合控制若干关键技术的发展及研究现状

1.2.1　相对运动动力学建模技术

相对运动动力学建模与分析是研究航天器近距离相对轨道和姿态运动的理论基础。下面重点从相对轨道动力学和姿轨耦合动力学两个方面进行介绍。

1.2.1.1 相对轨道动力学

目前，相对轨道描述方法主要分为代数法（动力学方法）和几何法（运动学方法）。代数法是以笛卡儿坐标描述的相对轨道动力学模型为基础，故也称为坐标法，便于进行相对运动制导与控制，由此得到的近似模型如著名的C-W方程[19]、T-H方程[20]等在相对轨道控制方面应用广泛。几何法是以轨道要素表示的相对运动模型为基础，故也称为轨道要素法，该法便于进行摄动分析，常用于轨道设计和构型保持[21-24]。两者精度相当，仅是在推导方法和使用上存有差异，不过导航输出参数大多为航天器相对位置和速度，因此前者更适合进行导航控制一体化设计。

经过半个多世纪的发展，C-W方程因其具有形式简洁且满足一定精度要求的解析解，已在近圆轨道交会[25-33]、编队飞行[34-36]等方面得到了广泛应用。C-W方程是基于球形中心引力体和圆参考轨道假设推导得到的，因此线性化误差和J_2项摄动是其主要误差来源[34]。为了进一步消除圆参考轨道的限制，Tschauner和Hempel[20]通过变量替换推导了以真近点角为自变量的线性化相对运动方程，即T-H方程，该方程适用于椭圆参考轨道的相对运动，但依然存在线性化误差和摄动误差。针对T-H方程解析解的求解问题，国内外学者展开了大量的研究[20,37-41]，但求解得到的状态转移矩阵大多形式比较复杂，不便于工程应用。2002年，Yamanaka和Ankersen[42]结合坐标变换给出了形式相对简单且适用于任意偏心率大小的椭圆轨道状态转移阵，但该方法仅提供了轨道面内角域形式的表达式，在应用时还需要进行多次角域和时域的转换。在此基础上，张刚和周荻[43]进一步推导了轨道面内与目标航天器初始状态和追踪时间相关的状态转移矩阵。2017年，我国青年学者Dang利用积分拆解组合技术，巧妙引入一种新型数学积分，首次获得了一组适用于所有开普勒轨道类型的无奇异完整解析解，系统性地解决了T-H方程的求解问题[44]。近年来，他将人工智能方法引入航天器轨道动力学，利用神经网络、强化学习等人工智能方法解决了一大类经典问题，取得了意想不到的优良效果，使轨道动力学表现出了“智能化”的特点[45]。

1.2.1.2 姿轨耦合动力学

两航天器间的相对姿态动力学可看作姿态跟踪问题，即航天器保持对运动目标的定向，该模型在空间交会对接、近距离观测和空间机器人抓捕等任务中经常用到。目前，研究的热点主要集中在六自由度姿轨耦合动力学问题上[46-53]，尤其是针对一些新型空间任务，需要考虑姿态运动对轨道运动的影响。例如，Lisano针对太阳帆推进航天器[48]，Somenzi等针对绳系航天器[50]，Stansbery和Cloutier[46]、Pan和Kapila[47]，Gaulocher[49]、Kristiansen等[52]针对航天器编队飞行任务，均考虑了轨道运动与姿态运动之间的耦合影响，建立了六自由度航天器姿轨耦合动力学模型。Segal 和 Gurfil 进一步考虑了相对姿态角速度对相对平动的运动学耦合作用，建立了两航天器上任意点间相对运动的耦合动力学模型，从而解决了点质量模型引起的偏差[54]。

然而，上述方法在建模时仍然是将轨道和姿态分开处理，这种分而治之的方法直接影响了相对位置和姿态的控制精度，并且增加了系统复杂性。为了建立航天器姿轨一体化动力学模型，需要寻找一种有力的数学工具能够统一描述航天器轨道运动和姿态运动。Junkins 和 Turner[55]、Chelnokov[56]、Waldvogel[57]等开展了利用四元数描述航天器轨道参数的研究，实现了轨道和姿态描述参数上的统一，这些方法具有一定的里程碑意义。但四元数仅有三个自由度，不能完全表征航天器的一般性空间运动，因此仍然没有实现航天器轨道和姿态动力学模型的完全统一。为了克服这个困难，Ploen 等[58]和 Sinclair 等[59]尝试了采用其他数学工具进行航天器姿轨耦合动力学的建模。

目前，对偶四元数代数因其能统一描述一般性刚体运动（包括旋转和平移）而备受青睐，已成功应用在机械[60-61]、控制[62]、捷联惯导[63-65]、空间机器人[66]、计算机辅助设计[67]等多个领域中分析和处理运动学与动力学问题。对偶代数由 Clifford[68]于 19 世纪率先提出，并经 Kotelnikov[69]和 Study[70]发展应用到机械领域。相比其他数学工具，如齐次变换、Q/T 法、李代数等，对偶四元数代数被证实是最简捷有效描述刚体一般性运动的方法[71-73]。Kotelnikov 转移定理表明：相对于定点的刚体运动学的所有矢量代数运算法则在自由刚体运动学的旋量代数中同样成立。因此，对偶四元数代数完全继承了四元数的性质，所建立的刚体一般性运动的动力学模型[74-75]也与利用四元数描述的刚体旋转运动的动力学模型具有相似的形式，可以借鉴其成熟理论进行相关研究。近年来，Pennestrì 等[76-77]进一步将经典的线性代数运算法则扩展到对偶代数中，并利用对偶代数完成了对人的肢体运动的运动学分析。张洪珠[78]、王剑颖等[79-82]、Filipe 和 Tsiotras[83-86]等相继开展了基于对偶代数的航天器姿轨一体化动力学建模与控制方法的研究，取得了一些卓有成效的成果。

1.2.2　视觉相对导航技术

目前，在近距离操作段，航天器视觉测量系统被广泛用来测量航天器间的相对位置和姿态参数。视觉测量系统具有重量轻、体积小、功耗低、集成度高以及隐蔽性好等优点，且测量精度随着相对距离的减小越来越高。根据相机数目和获取目标信息的方法，可以将视觉测量方法分为单目视觉、双目立体视觉以及多目立体视觉。单目视觉系统因其结构简单、易于标定，且不需要进行复杂的特征匹配等优点已被广泛应用于摄影测量、机器人导航和航天工程领域。它的缺点主要在于将 3D 物体投影到 2D 平面上会丢失深度信息，并导致所得到的映射关系是非线性的。基于特征点的单目视觉相对位姿确定问题是一个经典的 n 点透视（Perspective N Points，PnP）投影问题，自 20 世纪 80 年代以来已有众多国内外学者给出了大量的 PnP 问题解法[87]，这些算法大致可分为直接算法和迭代算法两大类。直接算法是采用代数方法直接求出相对位置和姿态参数，针对 P3P、P4P 以及 P5P 等问题，目前已经推导出多种解析算法[88-90]，这类算法具有计算量小、运算速度快等优点，但易受误差因素影响且精度不高，一般可用于迭代算法的初值计算。迭代算法是将 PnP 问题表示成受约束的非线性优化问题，然后采用数值迭代算法对其进行求解，进而求得相对位姿参数。这类算法[91-94]可降低成像误差噪声的影响，从而提高相对位姿确定精度，但计算量较大且算法的收敛性能受初始值的影响较大。

然而，单纯依赖视觉测量方法仅利用静态的几何关系进行相对位姿的解算，在实际应用过程中易受测量误差、量化误差、特征点提取误差以及可能的特征点匹配错误等误差因素影响，系统可靠性低，且计算精度与先验信息、迭代初值、迭代次数等因素均有关系。结合航天器动力学模型和惯性敏感器件，从状态估计的角度出发，可设计绝对导航[95-97]和相对导航滤波器[98-102]，研究表明这两种导航系统在相对位置和速度的估计以及相对轨道控制偏差上效果差不多[103]。绝对导航系统在惯性参考系下建立两航天器的绝对运动方程并进行状态估计，通过做差求解两航天器的相对状态参数；相对导航系统主要基于线性化的相对运动方程，状态维数较前者少，但不能求解出绝对运动参数。对于近距离相对运动而言，相对导航系统因其具有足够的精度和更高的计算效率常常被采用。

一般来说，标志点位置矢量定义在目标航天器（主星）的本体系中，而主从星之间的相对位置矢量常定义在主星的当地轨道坐标（Local-Vertical-Local-Horizontal，LVLH）系中。因此，为了便于构造视线测量量，参考文献[99-101]做了主星本体系与其 LVLH 系重合这一简化假设，进而估计出两航天器间的相对姿态。为了消除这一假设条件限制，Zhang 等[104-106]利用两航天器本体系相对于主星 LVLH 系的姿态四元数重新构造视线观测量，并先后提出了基于扩展卡尔曼滤波（Extended Kalman Filter，EKF）、无味卡尔曼滤波（Unscented Kalman Filter，UKF）以及容积卡尔曼滤波（Cubature Kalman Filter，CKF）的航天器间相对位姿确定算法。另外一种做法类似于参考文献

[102]，将相对运动方程描述在主星本体系而非 LVLH 系中，从而使得标志点位置矢量和相对位置矢量描述在同一坐标系里。但这样带来的缺点是导致相对运动方程更加复杂，另外由于主星的体系角速度相比轨道角速度变化更快且受惯性器件测量误差影响大，会给相对运动方程计算带来较大误差。

近年来，故障、失效或任务结束而丢弃在太空的航天器数量逐年增多，这些航天器不仅占用了宝贵的空间轨道资源，还可能危及其他航天器的安全，因此各国正在加紧开展面向空间非合作航天器的在轨服务技术研究[2-7,13,107]。要实现在轨捕获与维修，必须解决非合作航天器间的相对位姿测量问题。针对“哈勃”太空望远镜的在轨维修任务，Thienel 和 Sanner[108]通过估计惯性角动量，提出了一种非线性估计方法用于确定非合作航天器的角速度和姿态，但该方法只考虑了两航天器间的相对转动问题，且没有考虑敏感器测量噪声和惯性参数不确定性等影响。基于激光立体视觉系统，Aghili 和 Parsa[109]设计了一个自适应卡尔曼滤波器用于估计失控翻滚航天器的相对位姿，该方法实现了目标航天器（主星）惯性参数的在线估计。受这篇文章启发，Zhang 等[110-111]针对失控翻滚航天器进一步设计了基于单目视觉的相对导航和绝对导航滤波器。由于针对完全非合作目标的相对位姿确定技术具有重要的军事用途，目前公开的参考文献仍然较少。据报道，加拿大研发了基于模型匹配的双目视觉系统，该系统不需要目标三维结构参考模型的支持，能够处理完全非合作目标[112]。国内学者张世杰等[113]和徐文福等[114]也进行了相关工作的跟进研究。

1.2.3　近距离姿轨耦合控制技术

面对新型空间任务的高精度需求，传统的航天器轨道和姿态分别独立的控制方式存在着一定的局限性，尤其是在近距离操作段，相对位置和姿态的耦合控制问题逐渐受到众多学者的关注。目前，对姿轨耦合控制问题的研究方法主要分为两种：一种是先独立设计轨道控制器和姿态控制器，然后针对姿轨耦合部分做补偿与修正[115]；另一种是建立姿轨耦合动力学模型，设计一体化控制算法[46-47,116-122]。后者可以借鉴许多现有的控制器设计方法，并且能够提供完备的 Lyapunov 稳定性分析，因此较前者更为常用。

针对失控翻滚目标，Terui[116]，Stansbery 和 Cloutier[46]，Xin 等[117]分别采用滑模控制、状态独立黎卡提方程（State-Dependent Riccati Equation，SDRE）、θ-D 次优控制方法对航天器近距离自主交会的六自由度控制问题进行了研究。然而，他们的动力学模型中并没有体现出轨道和姿态之间的耦合作用，仅仅是轨道和姿态控制问题的简单叠加。针对航天器编队飞行控制问题，Pan 和

Kapila[47]考虑了由重力梯度力矩引起的相对位置和姿态耦合问题，推导了从星相对于主星的位置和姿态耦合动力学模型，并设计了自适应的非线性控制器；Wong 等[118]进一步针对无线速度和角速度测量信息的情形，通过设计一个高通滤波器，设计了非线性的输出反馈控制律；Kristiansen 等[52]利用 Backstepping 法构造了能保证系统的一致渐近稳定性的姿轨耦合控制算法，并将其与滑模和比例微分方法做了比较。国内学者吴云华等[119]利用高斯伪谱法，研究了采用单个连续小推力推力器以及反作用飞轮作为执行机构的编队卫星相对轨道和姿态耦合控制问题。卢伟等[120]同时考虑了控制输入和控制指令耦合问题，并利用反馈线性化理论设计了相对位置和姿态的一体化耦合控制算法。唐生勇[121]根据任务控制能力要求和系统约束条件，设计了一套冗余推力器系统实现了姿轨一体化控制。

近年来，众多学者开展了基于对偶代数的航天器姿轨耦合控制问题研究。相比以往轨道和姿态分别建模的情形，基于对偶代数的航天器姿轨耦合模型形式更加简洁、明了，计算效率更高。因此，现有成熟的控制理论，如比例微分（Proportional Differential，PD）控制、滑模变结构控制（Variable Structure Control，VSC）[79]、线性二次型调节器（Linear Quadratic Regulator，LQR）控制和有限时间稳定控制[80,123]等均被应用来实现航天器姿轨一体化控制，并取得了若干成果。鉴于该模型与传统的欧拉姿态动力学模型在形式上的相似性，Filipe 和 Tsiotras 借鉴已有的姿态运动控制律设计的方法，先后研究了无线速度和角速度测量信息反馈[83-84]、无先验质量和惯量参数信息[85-86]的航天器姿轨耦合控制问题，并提供了较为严格的 Lyapunov 稳定性证明。

1.3 本书主要思路及内容

本书采用理论分析、数学建模和计算机数值仿真相结合的方法，深入研究了椭圆轨道近距离相对导航与姿轨一体化控制方法，重在基本理论和方法的探讨，所提出的方法均适用于目标航天器（主星）轨道为椭圆或近圆轨道情形。各章的具体内容安排如下：

第 1 章是概论。首先介绍了本书的研究背景、目的和意义；其次从相对运动动力学建模、视觉相对导航方法以及近距离姿轨耦合控制三个方面进行了国内外研究进展综述；最后介绍了内容安排。

第 2 章是数学模型介绍。首先介绍了航天器相对轨道动力学模型，推导了相对姿态运动学方程；其次分别介绍了敏感器模型和执行机构模型，其中敏感器模型包含了视觉导航（Vision-based Navigation，VISNAV）系统的两种观测模型和陀螺测量模型，执行机构模型包含了喷气推力器和控制力矩陀螺模型。

第 3 章是基于非线性最小二乘法的航天器相对位姿确定方法研究。基于前述的视觉导航系统观测模型 1，本章采用非线性最小二乘法求解航天器相对位置和姿态，为后续设计的导航滤波器提供初值。综合考虑了标志点位置误差噪声和视觉导航系统测量噪声两方面影响，推导了观测误差噪声阵并将其用于非线性最小二乘算法中。数值仿真分为静态仿真和动态仿真两部分，验证了算法的有效性和可行性。仿真结果表明，标志点位置误差噪声对相对位置和姿态测量精度影响较大；随着主从星之间距离的接近，系统的可观测性越来越好，相对位置和姿态估计精度也越来越高。

第 4 章是基于状态估计的近距离相对导航方法研究。本章从状态估计的角度出发，结合航天器相对动力学设计导航滤波器，实现两航天器间相对位置、速度和姿态的估计。首先，基于视觉导航系统观测模型 2，提出了一种新的基于 EKF 的航天器间相对位姿确定算法。与传统算法相比，该算法额外估计了两航天器体系相对于主星 LVLH 系的姿态，从而不必要去做主星本体系与 LVLH 系重合这一假设，其实质是不需要先验的主星绝对姿态信息。其次，在此基础上进一步提出了基于 CKF 的航天器间相对位姿确定算法，并利用传播的四元数容积点集的加权平均值作为参考四元数，对所提的 CKF 导航滤波器进行了改进，在一定程度上提高了滤波性能。最后，针对主星失控翻滚情形，推导了在缺少主星惯性参数和角速度先验信息条件下的相对位姿确定算法。蒙特卡洛仿真表明，观测标志点数目达到 3 个及以上时，所设计的滤波器均能准确地估计相对位置、速度、姿态、主星角速度以及惯量比矢量。随着观测标志点数目的增加，估计精度也随之提高。

第 5 章是基于对偶代数的航天器姿轨一体化控制方法研究。本章建立了基于对偶代数的航天器轨/姿运动学和动力学模型，并在此基础上推导了基于对偶代数的姿轨耦合误差动力学方程，证明了其与传统的相对转动和相对平动误差动力学方程一致，但形式更加简洁、明了，计算效率更高。另外，还研究了基于对偶代数的航天器姿轨一体化控制方法，设计了“前馈补偿+误差 PD”复合控制律和全推力器控制系统。结合第 4 章中的相对导航方法给出了如何求解姿轨一体化控制律中所需参数导航估计值的方法，可进行导航控制一体化仿真。仿真结果表明设计的姿轨一体化控制算法是有效的，并且对质量和转动惯量的不确定性以及外界扰动具有较好的鲁棒性。

第 6 章是空间高精度姿态机动控制方法研究。本章主要分析了两种空间高精度姿态机动控制方法：一种是基于路径规划和 LQR 反馈控制相结合的姿态机动控制方法；另一种是模型误差预测控制方法。针对采用单框架控制力矩陀螺系统作为执行机构的航天器大角度姿态机动任务，提出了一种基于路径规划和 LQR 反馈控制相结合的姿态机动控制方法。针对固定时间能量最优和准时间最优两种情形，采用 Radau 伪谱法优化机动路径并回避奇异状态，将最优控制量作为参考输入，并利用基于 LQR 的最优反馈控制方法消除初始偏差、模型不确定性以及外界扰动等影响。针对参数不确定性以及外界干扰等情形，进一步分析了模型误差预测控制方法。仿真结果表明，在滑模变结构控制方法的基础上，采用模型误差预测控制算法可以有效地对参数模型误差和外界扰动引起的误差进行补偿。

参考文献

[1] WALTZ D M. On-orbit servicing of space systems [M]. Florida: Krieger Publishing Company, 1993.

[2] RUMFORD T E. Demonstration of Autonomous Rendezvous Technology (DART) Project Summary [C]//Proceedings of SPIE-The International Society for Optical Engineering, August 2003. USA: Space Systems Technology and Operations, 2003: 10-19.

[3] BARNHART D A, HUNTER R C, WESTON A R, et al. XSS-10 micro-satellite demonstration [R]. USA: AIAA Inc, 1998: 339-346.

[4] MITCHELL I T, GORTON T B, TASKOV K, et al. GN&C development of the xss-11 micro-satellite for autonomous rendezvous and proximity operations [C]//The 29th Annual AAS Guidance and Control Conference. USA: AIAA Inc, 2006: 33-42.

[5] WEISMULLER T, LEINZ M. GN&C demonstrated by the orbital express autonomous rendezvous and capture sensor system [C]//The 29th Annual AAS Guidance and Control Conference. USA: AIAA Inc, 2006: 54-60.

[6] EVANS J W, PINON Ⅲ F, MULDER T A. Autonomous rendezvous guidance and navigation for orbital express and beyond [C]//The 29th Annual AAS Guidance and Control Conference. USA: AIAA Inc, 2006: 1565-1574.

[7] Spacecraft for the universal modification of orbits (SUMO) program [EB/OL]. [2005-10-8]. http:// www.spaceref.ca/news/viewsr.html?pid=18321.

[8] CERESETTI A, TAMBURINI P. Automated transfer vehicle-a European versatile multi-mission vehicle [C]//AIAA Space Programs and Technologies

Conference, Huntsville, September 27-29, 1994. USA: AIAA Inc, 1994: 46-54.

[9] PERROTON G, BUSSON M. Enhancement of the European ATV for logistics missions to ISS [C]//AIAA Space Technology Conference, New Mexico, September 28-30, 1999. USA: AIAA Inc, 1999: 325-332.

[10] KAWANO I, MOKUNO M, KASAI T, et al. Result of autonomous rendezvous docking experiment of engineering test satellite-VII [J]. Journal of Spacecraft and Rockets, 2001, 38 (1): 105-111.

[11] KAWANO I, MOKUNO M, HORIGUCHI H, et al. In-orbit demonstration of an unmanned automatic rendezvous and docking system by the Japanese engineering test satellite ETS-VII [C]//AIAA Guidance, Navigation and Control Conference, August 1-3, 1994. USA: AIAA Inc, 1994: 950-960.

[12] MATSUMOTO T. Development of the proximity communication system (PROX) on ISS for H-II transfer vehicle (HTV) rendezvous and proximity operation [C]//21st International Communications Satellite Systems Conference, Yokohama, April 15-19, 2003. USA: AIAA Inc, 2003: 1-7.

[13] BAARD K. PRISMA demonstration mission for advanced rendezvous and formation flying technologies and sensors [J]. Nordicspace, 2005, 13 (4): 14-19.

[14] 周建平. 天宫一号/神舟八号交会对接任务总体评述 [J]. 载人航天, 2012, 18 (1): 1-5.

[15] 周建平. 我国空间站工程总体构想 [J]. 载人航天, 2013, 19 (2): 1-10.

[16] 王大铁, 胡启阳, 胡海东, 等. 非合作航天器自主相对导航研究综述 [J]. 控制理论与应用, 2018, 35 (10): 1392-1404.

[17] SOTTO E D, BASTANTE J C, DRAI R. System and GNC concept for rendezvous into ellipitical orbit for mars sample return misson [C]//AIAA Guidance, Navigation, and Control Conference and Exhibit, August 20-23, 2007. USA: AIAA Inc, 2007: 1-18.

[18] 马婷婷, 魏晨曦. 空间交会对接测量技术的发展 [J]. 中国航天, 2004 (7): 30-34.

[19] CLOHESSY W H, WILTSHIRE R S. Terminal guidance system for satellite rendezvous [J]. Journal of the Astronautical Sciences, 1960, 27 (9): 653-678.

[20] TSCHAUNER J, HEMPEL P. Rendezvous zu einem in elliptischer bahn umlaufenden ziel [J]. Acta Astronautica, 1965, 11 (2): 104-109.

[21] SENGUPTA P. Satellite relative motion propagation and control in the presence of J2 perturbations [D]. College Station, T X: Texas A&M University, 2003.

[22] 王虎妹，杨卫，李俊峰. 椭圆参照轨道的理想卫星编队队形设计 [J]. 清华大学报（自然科学版），2006，46（2）：266-270.

[23] 王虎妹，李俊峰，杨卫. 利用参照轨道要素研究卫星星座/编队的几何特征 [J]. 工程力学，2006，23（11）：177-181.

[24] JIANG F H, LI J F, BAOYIN H. Approximate analysis for relative motion of satellite formation flying in elliptical orbits [J]. Celestial Mechanics and Dynamical Astronomy, 2007, 98 (1): 31-66.

[25] PRUSSING J E. Optimal multiple-impulse orbital rendezvous [D]. Cambridge: Massachusetts Institute of Technology, 1967.

[26] PRUSSING J E. Optimal two-and three-impulse fixed-time rendezvous in the vicinity of a circular orbit [J]. AIAA Journal, 1970, 8 (7): 1221-1228.

[27] PRUSSING J E, CHIU J H. Optimal multiple-impulse time-fixed rendezvous between circular orbit [J]. Journal of Guidance, Control and Dynamics, 1986, 9 (1): 17-22.

[28] JEZEWSKI D J, DONALDSON J D. An analytic approach to optimal rendezvous using clohessy-wiltshire equations [J]. Journal of the Astronautical Sciences, 1979, 27 (3): 293-310.

[29] JEZEWSKI D. Primer vector theory applied to the linear relative-motion equations [J]. Optimal Control Applications and Methods, 1980, 1 (4): 387-401.

[30] CARTER T E. Fuel-optimal maneuvers of spacecraft relative to a point in circular orbit [J]. Journal of Guidance, Control and Dynamics, 1984, 7 (6): 710-716.

[31] CARTER T E, ALVAREZ S A. Four-impulse rendezvous near circular orbit [C]//AIAA/AAS Astrodynamics Specialist Conference and Exhibit, Boston, August 10-12, 1998. USA: AIAA Inc, 1998: 1-19.

[32] CARTER T E, ALVAREZ S A. Quadratic-based computation of four-impulse optimal rendezvous near circular orbit [J]. Journal of Guidance,

Control and Dynamics, 2000, 23 (1): 109-117.

[33] LUO Y Z, ZHANG J, LI H Y, et al. Interactive optimization approach for optimal impulsive rendezvous using primer vector and evolutionary algorithms [J]. Acta Astronautica, 2010, 67 (3/4): 396-405.

[34] SPARKS A. Satellites formationkeeping control in the presence of gravity perturbations [C]//Proceedings of the American Control Conference, Chicago, Illinois, June 2000. USA: IEEE Inc, 2000: 844-848.

[35] ALFRIEND K T, SCHAUB H, GIM D W. Gravitational perturbations nonlinearity and circular orbit assumption effects on formation flying control strategies [J]. Advances in the Astronautical Sciences, 2000, 104: 139-158.

[36] 傅敬博．考虑性能约束的 SAR 卫星编队姿态协同控制 [D]．哈尔滨：哈尔滨工业大学，2021.

[37] LAWDEN D F. Optimal trajectories for space navigation [M]. London: Butterworths, 1963.

[38] WOLFSBERGER W, WEI B J, RANGNITT D. Strategies and schemes for rendezvous on geostationary transfer orbit [J]. Acta Astronautica, 1983, 10 (8): 527-538.

[39] CARTER T E. State transition matrices for terminal rendezvous studies: brief survey and new example [J]. Journal of Guidance, Control, and Dynamics, 1998, 21 (1): 148-155.

[40] MELTON R G. Time-explicit representation of relative motion between elliptical orbits [J]. Journal of Guidance, Control, and Dynamics, 2000, 23 (4): 604-610.

[41] GIM D W, ALFRIEND K T. State transition matrix of relative motion for the perturbed noncircular reference orbit [J]. Journal of Guidance, Control, and Dynamics, 2002, 26 (6): 956-971.

[42] YAMANAKA K J, ANKERSEN F. New state transition matrix for relative motion on an arbitrary elliptical orbit [J]. Journal of Guidance, Control, and Dynamics, 2002, 25 (1): 60-66.

[43] 张刚，周荻．交会椭圆轨道目标 Hill 制导偏差分析 [J]．宇航学报，2010，31 (3)：707-713.

[44] DANG Z H. Solutions of Tschauner-Hempel equations [J]. Journal of Guidance, Control and Dynamics, 2017, 40 (11): 2956-2960.

[45] ZHENG M Z, LUO J J, DANG Z H. Feedforward neural network based time-

varying state - transition - matrix of Tschauner - Hempel equations [J]. Advances in Space Research, 2022, 69 (2): 1000-1011.

[46] STANSBERY D T, CLOUTIER J R. Position and attitude control of a spacecraft using the state-dependent Riccati equation technique [C]//Proceedings of the American Control Conference, Chicago, Illinois, June 2000. USA: IEEE Inc, 2000: 1867-1871.

[47] PAN H Z, KAPILA V. Adaptive nonlinear control for spacecraft formation flying with coupled translational and attitude dynamics [C]//Proceedings of the 40th IEEE Conference on Decision and Control, Orlando, Florida, December 2001. USA: IEEE Inc, 2001: 2057-2062.

[48] LISANO M E. A practical six-degree-of-freedom solar sail dynamics model for optimizing solar sail trajectories with torque constraints [C]//AIAA Guidance, Navigation, and Control Conference and Exhibit, Rhode Island, 2004. USA: AIAA Inc, 2004: 828-838.

[49] GAULOCHER S. Modeling the coupled translational and rotational relative dynamics for formation flying control [C]//AIAA Guidance, Navigation, and Control Conference and Exhibit, San Francisco, August 2005. USA: AIAA Inc, 2005: 1-6.

[50] SOMENZI L, IESS L, PELAEZ J. Linear stability analysis of electrodynamic thethers [J]. Journal of Guidance, Control, and Dynamics, 2005, 28 (5): 843-849.

[51] SUBBARAO K, WELSH S. Nonlinear control of motion sychronization for satellite proximity operations [J]. Journal of Guidance, Control, and Dynamics, 2008, 31 (5): 1284-1294.

[52] KRISTIANSEN R, NICKLASSON P J, GRAVDAHL J T. Spacecraft coordination control in 6DOF: Integrator backstepping vs passivity-based control [J]. Automatica, 2008, 44 (11): 2896-2901.

[53] XIN M, PAN H J. Integrated nonlinear optimal control of spacecraft in proximity operations [J]. International Journal of Control, 2010, 83 (2): 347-363.

[54] SEGAL S, GURFIL P. Effect of kinematic rotation-translation coupling on relative spacecraft translational dynamics [J]. Journal of Guidance, Control and Dynamics, 2009, 32 (2): 1045-1050.

[55] JUNKINS J L, TURNER J D. On the analogy between orbital dynamics and rigid body dynamics [J]. Journal of the Astronautical Sciences, 1979, 27

(4): 345-358.
[56] CHELNOKOV Y N. The use of quaternions in the optimal control problems of motion of the center of mass of a spacecraft in a newtonian gravitational field: I [J]. Cosmic Research, 2001, 39 (5): 470-484.
[57] WALDVOGEL J. Quaternions and the perturbed kepler problem [J]. Celestial Mechanics and Dynamical Astronomy, 2006, 95 (1-4): 201-212.
[58] PLOEN S R, HADAEGH F Y, SCHARF D P. Rigid body equations of motion for modeling and control of spacecraft formations-part I: absolute equations of motion [C]//Proceeding of the 2004 American Control Conference, Boston, 2004. USA: IEEE Inc, 2004: 3646-3653.
[59] SINCLAIR A J. HURTADO J E, JUNKINS J L. Application of the cayley form to general spacecraft motion [J]. Journal of Guidance, Control, and Dynamics, 2006, 29 (2): 368-373.
[60] YANG A T. Application of quaternion algebra and dual numbers to the analysis of spatial mechanisms [D]. New York: Columbia University, 1964.
[61] DOOLEY J R, MCCARTHY J M. Spatial rigid body dynamics using dual quaternion components [C]//Proceedings of the 1991 IEEE International Conference on Robotics and Automation Sacramento, California, 1991. USA: IEEE Inc, 1991: 90-95.
[62] 韩大鹏. 基于四元数代数和李群框架的任务空间控制方法研究 [D]. 长沙: 国防科学技术大学, 2008.
[63] BRANETS V N, SHMYGLEVSKY I P. Introduction to the Theory of Strapdown Inertial Navigation System [M]. Moscow: Nauka Inc, 1992.
[64] WU Y X, HU X P, HU D W, et al. Strapdown inertial navigation system algorithms based on dual quaternions [J]. IEEE Aerospace and Electronic Systems, 2005, 41 (1): 110-132.
[65] 武元新. 对偶四元数导航算法与非线性高斯滤波研究 [D]. 长沙: 国防科学技术大学, 2005.
[66] DANIILIDIS K. Hand-eye calibration using dual quaternions [J]. International Journal of Robotics Research, 1999, 18 (3): 286-298.
[67] ROONEY J. A comparison of representation of general spatial screw displacement [J]. Environment and Planning B, 1978, 5 (1): 45-88.
[68] CLIFFORD W. Preliminary sketch of biquaternions [J]. Proceedings of the

London Mathematical Society, 1873, s1/4 (1): 381-395.
[69] KOTELNIKOV A P. Screw calculus and some applications to geometry and mechanics [D]. Kazan: Annuals of Imperial University of Kazan, 1895.
[70] STUDY E. Geometrie der dynamen [M]. Leipzig: BG Teubner, 1903.
[71] FUNDA J, PAUL R P. A computational analysis of screw transformations in robotics [J]. IEEE Transactions on Robotics and Automation, 1990, 6 (3): 348-356.
[72] FUNDA J, TAYLOR R H, PAUL R P. On homogeneous transformations, quaternions, and computational efficiency [J]. IEEE Transactions on Robotics and Automation, 1990, 6 (3): 382-388.
[73] ASPRAGATHOS N A, DIMITROS J K. A comparative study of three methods for robot kinematics [J]. IEEE Transactions on Systems, Man and Cybernetics Part B: Cybernetics, 1998, 28 (2): 135-145.
[74] BRODSKY V, SHOHAM M. The dual inertia operator and its application to robot dynamics [J]. Journal of Mechanical Design, 1994, 116 (4): 1189-1195.
[75] BRODSKY V, SHOHAM M. Dual numbers representation of rigid body dynamics [J]. Mechanism and Machine Theory, 1999, 34 (5): 693-718.
[76] PENNESTRÌ E, STEFANELLI R. Linear algebra and numerical algorithms using dual numbers [J]. Multibody System Dynamics, 2007, 18 (3): 323-344.
[77] PENNESTRÌ E, VALENTINI P P. Dual quaternions as a tool for rigid body motion analysis: a tutorial with an application to biomechanics [C]//Multibody Dynamcs, Eccomas Thematic Conference, Poland, June 2009. USA: [s. n.]. 2009: 1-14.
[78] 张洪珠. 基于对偶四元数的航天器姿轨一体化动力学建模与控制 [D]. 哈尔滨：哈尔滨工业大学，2010.
[79] WANG J Y, SUN Z W. 6-DOF robust adaptive terminal sliding mode control for spacecraft formation flying [J]. Acta Astronautica, 2012, 73: 76-87.
[80] WANG J Y, LIANG H Z, SUN Z W, et al. Finite-time control for spacecraft formation with dual-number-based description [J]. Journal of Guidance, Control and Dynamics, 2012, 35 (3): 950-962.
[81] WANG J Y, LIANG H Z, SUN Z W, et al. Relative motion coupled control based on dual quaternion [J]. Aerospace Science and Technology, 2013,

25 (1): 102-113.

[82] 王剑颖. 航天器姿轨一体化动力学建模、控制与导航方法研究 [D]. 哈尔滨: 哈尔滨工业大学, 2013.

[83] FILIPE N, TSIOTRAS P. Rigid body motion tracking without linear and angular velocity feedback using dual quaternions [C]//IEEE European Control Conference, Piscataway. USA: IEEE Inc, 2013: 329-334.

[84] FILIPE N, TSIOTRAS P. Simultaneous position and attitude control without linear and angular velocity feedback using dual quaternions [C]//IEEE European Control Conference, Piscataway. USA: IEEE Inc, 2013: 4808-4813.

[85] FILIPE N, TSIOTRAS P. Adaptive model - independent tracking of rigid body position and attitude motion with mass and inertia matrix identification using dual quaternions [C]//AIAA Guidance, Navigation, and Control Conference, Boston, 2013. USA: AIAA Inc, 2003: 5805-5819.

[86] FILIPE N, TSIOTRAS P. Adaptive position and attitude-tracking controller for satellite proximity operations using dual quaternions [J]. Journal of Guidance, Control, and Dynamics, 2015, 38 (4): 566-577.

[87] WU Y H, HU Z Y. PnP problem revisited [J]. Journal of Mathematical Imaging and Vision, 2006, 24 (1): 131-141.

[88] GAO X S, HOU X R, TANG J L, et al. Complete solution classification for the perspective-three-point problem [J]. IEEE Transactions on Pattern Analysis and Machine Intelligence, 2003, 25 (8): 930-943.

[89] HU Z Y, WU C F. A note on the number of solutions of the noncoplanar P4P problem [J]. IEEE Transactions on Pattern Analysis and Machine Intelligence, 2002, 24 (4): 550-555.

[90] NISTER D. An efficient solution to the five-point relative pose problem [J]. IEEE Transactions on Pattern Analysis and Machine Intelligence, 2004, 26 (6): 756-770.

[91] LU C P, HAGER G D, MJOLSNESS E. Fast and globally convergent pose estimation from video images [J]. IEEE Transactions on Pattern Analysis and Machine Intelligence, 2000, 22 (6): 610-622.

[92] SCHWEIGHOFER G, PINZ A. Robust pose estimation from a planar target [J]. IEEE Transactions on Pattern Analysis and Machine Intelligence, 2006, 28 (12): 2024-2030.

[93] SCHWEIGHOFER G, PINZ A. Globally optimal O (n) solution to the PnP

problem for general camera models [C]//Proceedings of the British Machine Vision Conference 2008, Leeds, September 2008. London: British Machine Vision Association, 2008: 1-10.

[94] 张世杰，曹喜滨，张凡，等. 基于特征点的空间目标三维位姿单目视觉确定算法 [J]. 中国科学：信息科学，2010，40（4）：591-604.

[95] WODFFINDEN D C. GELLER D K. Relative angles-only navigation and pose estimation for autonomous orbital rendezvous [J]. Journal of Guidance, Control, and Dynamics, 2007, 30 (5): 1455-1469.

[96] SCHMIDT J, GELLER D K, CHAVEZ F. Viability of angles-only navigation for orbital rendezvous operation [C]//AIAA Guidance, Navigation, and Control Conference, Toronto, Aug 2010. USA: AIAA Inc, 2010: 1-25.

[97] HABLANI H B. Autonomous inertial relative navigation with sight-line-stabilized integrated sensors for spacecraft rendezvous [J]. Journal of Guidance, Control, and Dynamics, 2009, 32 (1): 172-183.

[98] HABLANI H B, TAPPER M L, DANA-BASHIAN D J. Guidance and relative navigation for autonomous rendezvous in a circular orbit [J]. Journal of Guidance, Control, and Dynamics, 2002, 25 (3): 553-562.

[99] ALONSO R, CRASSIDIS J, JUNKINS J. Vision-based relative navigation for formation flying of spacecraft [C]//AIAA Guidance, Navigation, and Control Conference, Denver, Aug 2000. USA: AIAA Inc, 2000: 1-11.

[100] KIM S G, CRASSIDIS J L, CHENG Y, et al. Kalman filtering for relative spacecraft attitude and position estimation [J]. Journal of Guidance, Control, and Dynamics, 2007, 30 (1): 133-143.

[101] XING Y J, CAO X B, ZHANG S J, et al. Relative position and attitude estimation for satellite formation with coupled translational and rotational dynamics [J]. Acta Astronautica, 2010, 67: 455-467.

[102] TANG X J, YAN J, ZHONG D D. Square-root sigma-point kalman filtering for spacecraft relative navigation [J]. Acta Astronautica, 2010, 66: 704-713.

[103] JENKINS S C, GELLER D K. State estimation and targeting for autonomous rendezvous and proximity operations [C]//The 30th Annual AAS Guidance and Control Conference. USA: AIAA Inc, 2007: 1071-1102.

[104] ZHANG L J, YANG H B, ZHANG S F, et al. Kalman filtering for relative spacecraft attitude and position estimation: a revisit [J]. Journal of Guid-

ance, Control and Dynamics, 2014, 37 (5): 1706-1711.

[105] ZHANG L J, LI T, YANG H B, et al. Unscented Kalman filtering for relative spacecraft attitude and position estimation [J]. The Journal of Navigation, 2015, 68 (3): 528-548.

[106] ZHANG L J, YANG H B, LU H P, et al. Cubature Kalman filtering for relative spacecraft attitude and position estimation [J]. Acta Astronautica, 2014, 105 (1): 254-264.

[107] AMICO S D', ARDANS J S, GAIAS G, et al. Noncooperative rendezvous using angles-only optical navigation: system design and flight results [J]. Journal of Guidance, Control and Dynamics, 2013, 36 (6): 1576-1595.

[108] THIENEL J K, SANNER R M. Hubble space telescope angular velocity estimation during the robotic servicing mission [J]. Journal of Guidance, Control, and Dynamics, 2007, 30 (1): 29-34.

[109] AGHILI F, PARSA K. Motion and parameter estimation of space objects using laser-vision data [J]. Journal of Guidance, Control, and Dynamics, 2009, 32 (2): 537-549.

[110] ZHANG L J, ZHANG S F, YANG H B, et al. Relative attitude and position estimation for a tumbling spacecraft [J]. Aerospace Science and Technology, 2015, 42: 97-105.

[111] ZHANG L J, QIAN S, ZHANG S F, et al. Research on angles-only SINS/CNS relative position and attitude determination algorithm for a tumbling spacecraft [J]. Proc. IMechE Part G: Journal of Aerospace Engineering, 2017, 231 (2): 218-228.

[112] BAI W C, SHI L L, XIAO X L, et al. Pose measurement and tracking of non-cooperative satellite based on stereo vision [C]//Proceedings of the 40th Chinese Control Conference, July 2021. USA: IEEE Computer Society, 2021: 8234-8240.

[113] 张世杰，曹喜滨，陈闽．非合作航天器间相对位姿的单目视觉确定算法［J］．南京理工大学学报，2006，30（5）：564-568.

[114] 徐文福，刘宇，梁斌，等．非合作航天器的相对位姿测量［J］．光学精密工程，2009，17（7）：1570-1581.

[115] LENNOX S E. Coupled orbital and attitude control simulations for spacecraft formation flying [C]//The 2004 AIAA Region I-MA Student Conference. USA: AIAA Inc, 2004: 1-10.

[116] TERUI F. Position and attitude control of a spacecraft by sliding mode control [C]//Proceedings of the American Control Conference, Pennsylvania, June 1998. USA: IEEE Inc, 1998: 217-221.
[117] XIN M, BALAKRISHNAN S N, STANSBERY D T. Spacecraft position and attitude control with theta-d technique [C]//AIAA Aerospace Sciences Meeting and Exhibit. USA: AIAA Inc, 2004: 1-7.
[118] WONG H, PAN H Z, KAPILA V. Output feedback control for spacecraft formation flying with coupled translation and attitude dynamics [C]//Proceedings of 2005 American Control Conference, Portland, 2005. USA: IEEE Inc, 2005: 2419-2426.
[119] 吴云华，曹喜滨，张世杰，等. 编队卫星相对轨道与姿态一体化耦合控制 [J]. 南京航空航天大学学报，2010，42 (1): 13-20.
[120] 卢伟，耿云海，陈雪芹，等. 在轨服务航天器对目标的相对位置和姿态耦合控制 [J]. 航空学报，2011，32 (5): 857-865.
[121] 唐生勇. 航天器过驱动控制分配方法及应用研究 [D]. 哈尔滨：哈尔滨工业大学，2012.
[122] 彭智宏，穆京京，张力军，等. 基于对偶四元数的航天器相对位置和姿态耦合控制 [J]. 飞行器测控学报，2013，32 (6): 549-554.
[123] ZHANG F, DUAN G G. Robust integrated translation and rotation finite-time maneuver of a rigid spacecraft based on dual quaternion [C]//AIAA Guidance, Navigation, and Control Conference, Portland, 2011. USA: AIAA Inc, 2011: 1-17.

第 2 章　航天器相对动力学、敏感器与执行机构模型

本章将首先介绍椭圆轨道航天器相对轨道动力学模型，并推导相对姿态运动学方程；其次分别介绍敏感器模型和执行机构模型，其中敏感器模型包含 VISNAV 系统和陀螺测量模型，执行机构模型包含喷气推力器和控制力矩陀螺模型。

2.1　航天器相对轨道动力学模型

2.1.1　坐标系定义

1. 地心（赤道）惯性坐标系（简记为 I 系）

地心（赤道）惯性坐标系原点在地心处，x 轴在赤道面内指向平春分点，z 轴垂直于赤道平面指向北极，y 轴垂直于 x 轴和 z 轴，并构成右手法则。

2. 主星轨道坐标系（简记为 H 系）

主星轨道坐标系原点在主星质心，x 轴与主星的地心矢量重合，由地心指向航天器；y 轴在轨道平面内与 x 轴垂直并指向运动方向为正；z 轴由右手法则规定，沿轨道面法向并与动量矩矢量方向一致，如图 2.1 所示。

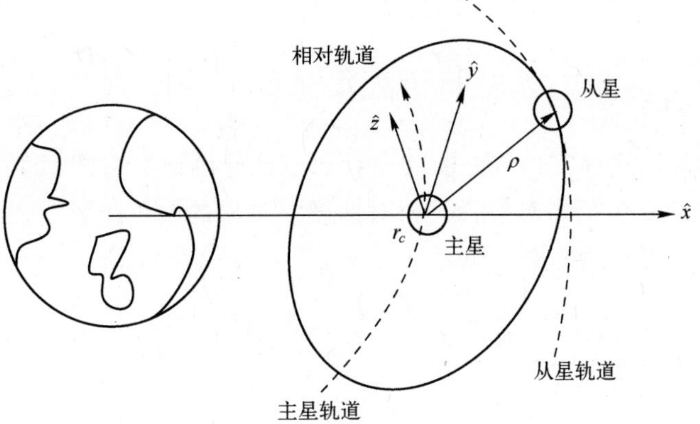

图 2.1　两航天器相对运动

3. 卫星本体坐标系

卫星本体坐标系原点在卫星质心，三轴固连在星体上且与卫星本体惯量主轴方向一致。主星（Chief）和从星（Deputy）的本体系分别记为 c 系和 d 系。

2.1.2 椭圆轨道相对动力学方程

不失一般性，考虑主星轨道为椭圆轨道情形，假设在主星 H 系中定义的相对轨道坐标 $\boldsymbol{\rho}\equiv[x,y,z]^{\mathrm{T}}$ 与主星地心矢径相比是小量，不考虑控制加速度，则椭圆轨道相对运动动力学方程（附录 A）可简化为[1]

$$\begin{cases}\ddot{x}-x\dot{\theta}^2\left(1+2\dfrac{r_c}{p}\right)-2\dot{\theta}\left(\dot{y}-y\dfrac{\dot{r}_c}{r_c}\right)=\boldsymbol{\varpi}_x\\ \ddot{y}+2\dot{\theta}\left(\dot{x}-x\dfrac{\dot{r}_c}{r_c}\right)-y\dot{\theta}^2\left(1-\dfrac{r_c}{p}\right)=\boldsymbol{\varpi}_y\\ \ddot{z}+z\dot{\theta}^2\dfrac{r_c}{p}=\boldsymbol{\varpi}_z\end{cases} \tag{2.1}$$

式中：p，r_c 和 $\dot{\theta}$ 分别为主星轨道的半通径、地心距和真近点角速率。扰动加速度矢量 $\boldsymbol{\varpi}\equiv[\boldsymbol{\varpi}_x\ \boldsymbol{\varpi}_y\ \boldsymbol{\varpi}_z]^{\mathrm{T}}$ 是零均值高斯白噪声过程，满足

$$\boldsymbol{E}\{\boldsymbol{\varpi}(t)\boldsymbol{\varpi}^{\mathrm{T}}(\tau)\}=\sigma_{\varpi}^2\delta(t-\tau)\boldsymbol{I}_{3\times3} \tag{2.2}$$

其中：$\boldsymbol{E}\{\cdot\}$ 表示期望；上标 T 表示转置；$\delta(t-\tau)$ 为 Diracδ 函数；$\boldsymbol{I}_{3\times3}$ 为 3×3 的单位阵；σ_{ϖ} 为扰动加速度的标准差。

对于主星的轨道运动，有

$$\ddot{\theta}=-2\frac{\dot{r}_c}{r_c}\dot{\theta},\quad \ddot{r}_c=r_c\dot{\theta}^2\left(1-\frac{r_c}{p}\right) \tag{2.3}$$

因此，利用式（2.1）和式（2.3）可以推导得到 10 维非线性状态空间模型。状态矢量 $\boldsymbol{x}_p$ 包括相对位置、相对速度、主星的轨道半径、径向速率、真近点角以及真近点角速率，即

$$\begin{aligned}\boldsymbol{x}_p&=[x\quad y\quad z\quad \dot{x}\quad \dot{y}\quad \dot{z}\quad r_c\quad \dot{r}_c\quad \theta\quad \dot{\theta}]^{\mathrm{T}}\\ &\equiv[x_{p1}\quad x_{p2}\quad x_{p3}\quad x_{p4}\quad x_{p5}\quad x_{p6}\quad x_{p7}\quad x_{p8}\quad x_{p9}\quad x_{p10}]^{\mathrm{T}}\end{aligned} \tag{2.4}$$

整理可得 10 维非线性微分方程组，即

$$
\dot{\boldsymbol{x}}_p=\boldsymbol{f}(\boldsymbol{x}_p)\equiv\begin{bmatrix}
x_{p_4}\\
x_{p_5}\\
x_{p_6}\\
x_{p_1}x_{p_{10}}^2(1+2x_{p_7}/p)+2x_{p_{10}}(x_{p_5}-x_{p_2}x_{p_8}/x_{p_7})\\
-2x_{p_{10}}(x_{p_4}-x_{p_1}x_{p_8}/x_{p_7})+x_{p_2}x_{p_{10}}^2(1-x_{p_7}/p)\\
-x_{p_7}x_{p_{10}}^2x_{p_3}/p\\
x_{p_8}\\
x_{p_7}x_{p_{10}}^2(1-x_{p_7}/p)\\
x_{p_{10}}\\
-2x_{p_8}x_{p_{10}}/x_{p_7}
\end{bmatrix} \tag{2.5}
$$

2.2　航天器相对姿态运动学模型

2.2.1　姿态四元数运动学方程

目前，最常用的姿态参数是姿态四元数[2]，其优点主要在于用其表示的姿态运动学方程为线性形式，计算量小，且不存在奇异性。四元数是一个四维矢量，定义为 $\boldsymbol{q}=[\boldsymbol{\varrho}^{\mathrm{T}}\quad q_4]^{\mathrm{T}}$，其中$\boldsymbol{\varrho}\equiv[q_1\quad q_2\quad q_3]^{\mathrm{T}}=\boldsymbol{e}\sin(\boldsymbol{\phi}/2)$，$q_4=\cos(\boldsymbol{\phi}/2)$，这里，$\boldsymbol{e}$ 为单位欧拉旋转轴，$\boldsymbol{\phi}$ 为旋转角。姿态矩阵可用四元数描述为[3]

$$
\begin{aligned}
\boldsymbol{A}(\boldsymbol{q})&=\begin{bmatrix}
q_1^2-q_2^2-q_3^2+q_4^2 & 2(q_1q_2+q_3q_4) & 2(q_1q_3-q_2q_4)\\
2(q_1q_2-q_3q_4) & -q_1^2+q_2^2-q_3^2+q_4^2 & 2(q_2q_3+q_1q_4)\\
2(q_1q_3+q_2q_4) & 2(q_2q_3-q_1q_4) & -q_1^2-q_2^2+q_3^2+q_4^2
\end{bmatrix}\\
&=(q_4^2-\|\boldsymbol{\varrho}\|^2)\boldsymbol{I}_{3\times3}+2\boldsymbol{\varrho}\boldsymbol{\varrho}^{\mathrm{T}}-2q_4[\boldsymbol{\varrho}\times]=\boldsymbol{\Xi}^{\mathrm{T}}(\boldsymbol{q})\boldsymbol{\Psi}(\boldsymbol{q})
\end{aligned} \tag{2.6}
$$

其中，

$$
\boldsymbol{\Xi}(\boldsymbol{q})\equiv\begin{bmatrix} q_4\boldsymbol{I}_{3\times3}+[\boldsymbol{\varrho}\times]\\ -\boldsymbol{\varrho}^{\mathrm{T}}\end{bmatrix},\quad \boldsymbol{\Psi}(\boldsymbol{q})\equiv\begin{bmatrix} q_4\boldsymbol{I}_{3\times3}-[\boldsymbol{\varrho}\times]\\ -\boldsymbol{\varrho}^{\mathrm{T}}\end{bmatrix} \tag{2.7}
$$

式中：$\boldsymbol{I}_{3\times3}$为 3×3 的单位阵；$[\boldsymbol{\varrho}\times]$为反对称阵，形式为

$$
[\boldsymbol{\varrho}\times]\equiv\begin{bmatrix} 0 & -q_3 & q_2\\ q_3 & 0 & -q_1\\ -q_2 & q_1 & 0\end{bmatrix} \tag{2.8}
$$

四元数的一个优点就是连续的姿态旋转可用四元数乘法进行描述，即 $\boldsymbol{A}(\boldsymbol{q}')\boldsymbol{A}(\boldsymbol{q})=\boldsymbol{A}(\boldsymbol{q}'\otimes\boldsymbol{q})$ 。值得注意的是，本书沿用了参考文献［4］中定义的四元数乘法规则，使得单位四元数乘法与姿态阵乘法顺序一致，与Hamiliton 建立的常用四元数乘法规则恰好相反[5]。这里，四元数乘法规则定义为

$$\boldsymbol{q}'\otimes\boldsymbol{q}\equiv\begin{bmatrix} q_4'\boldsymbol{\varrho}+q_4\boldsymbol{\varrho}'-\boldsymbol{\varrho}'\times\boldsymbol{\varrho} \\ q_4'q_4-\boldsymbol{\varrho}'\cdot\boldsymbol{\varrho} \end{bmatrix} \tag{2.9}$$

而传统的 Hamiliton 建立的四元数乘法规则为

$$\boldsymbol{q}'\otimes\boldsymbol{q}\equiv\begin{bmatrix} q_4'\boldsymbol{\varrho}+q_4\boldsymbol{\varrho}'+\boldsymbol{\varrho}'\times\boldsymbol{\varrho} \\ q_4'q_4-\boldsymbol{\varrho}'\cdot\boldsymbol{\varrho} \end{bmatrix} \tag{2.10}$$

实际上，四元数乘法具有双线性特性，且可用矩阵形式表示为

$$\boldsymbol{q}'\otimes\boldsymbol{q}=[\boldsymbol{\Psi}(\boldsymbol{q}')\quad \boldsymbol{q}']\boldsymbol{q}\equiv\{\boldsymbol{q}'\}_L\boldsymbol{q} \tag{2.11}$$

$$\boldsymbol{q}'\otimes\boldsymbol{q}=[\boldsymbol{\Xi}(\boldsymbol{q})\quad \boldsymbol{q}]\boldsymbol{q}'\equiv\{\boldsymbol{q}\}_R\boldsymbol{q}' \tag{2.12}$$

分析可知，$\{\boldsymbol{q}\}_L$，$\{\boldsymbol{q}\}_R$均为四阶正交矩阵，且满足$\{\boldsymbol{q}^{-1}\}_L=\{\boldsymbol{q}\}_L^{-1}=\{\boldsymbol{q}\}_L^{\mathrm{T}}$，$\{\boldsymbol{q}^{-1}\}_R=\{\boldsymbol{q}\}_R^{-1}=\{\boldsymbol{q}\}_R^{\mathrm{T}}$，进一步注意到

$$\{\boldsymbol{q}\}_L\{\boldsymbol{q}^{-1}\}_R=\{\boldsymbol{q}^{-1}\}_R\{\boldsymbol{q}\}_L=\begin{bmatrix} \boldsymbol{A}(\boldsymbol{q}) & \boldsymbol{0}_{3\times1} \\ \boldsymbol{0}_{3\times1}^{\mathrm{T}} & 1 \end{bmatrix} \tag{2.13}$$

那么，对于任意一个三维矢量 $\boldsymbol{v}$ ，可定义它的广义四元数为

$$\bar{\boldsymbol{v}}\equiv\begin{bmatrix} \boldsymbol{v} \\ 0 \end{bmatrix} \tag{2.14}$$

则根据式（2.13）可得

$$\bar{\boldsymbol{v}}'\equiv\begin{bmatrix} \boldsymbol{v}' \\ 0 \end{bmatrix}=\begin{bmatrix} \boldsymbol{A}(\boldsymbol{q})\boldsymbol{v} \\ 0 \end{bmatrix}=\boldsymbol{q}\otimes\bar{\boldsymbol{v}}\otimes\boldsymbol{q}^{-1} \tag{2.15}$$

因此，矢量的坐标旋转变成了四元数代数运算。另外，四元数满足正交性约束条件$\|\boldsymbol{q}\|=1$ ，其逆存在且等于其共轭，即有 $\boldsymbol{q}^{-1}=\boldsymbol{q}^{*}=[-\boldsymbol{\varrho}^{\mathrm{T}}\quad q_4]^{\mathrm{T}}$，满足 $\boldsymbol{q}\otimes\boldsymbol{q}^{-1}=\boldsymbol{q}^{-1}\otimes\boldsymbol{q}=[0\quad 0\quad 0\quad 1]^{\mathrm{T}}$。

姿态四元数运动学方程为[3]

$$\dot{\boldsymbol{q}}=\frac{1}{2}\boldsymbol{\Xi}(\boldsymbol{q})\boldsymbol{\omega}=\frac{1}{2}\boldsymbol{\Omega}(\boldsymbol{\omega})\boldsymbol{q} \tag{2.16}$$

式中：$\boldsymbol{\omega}$ 是三分量的角速度矢量，且有

$$\boldsymbol{\Omega}(\boldsymbol{\omega})\equiv\begin{bmatrix} -[\boldsymbol{\omega}\times] & \boldsymbol{\omega} \\ -\boldsymbol{\omega}^{\mathrm{T}} & 0 \end{bmatrix} \tag{2.17}$$

通常，在设计姿态控制律时，会将式（2.16）展开为

$$\begin{aligned}\dot{\boldsymbol{\varrho}}&=\frac{1}{2}([\boldsymbol{\varrho}\times]+q_4\boldsymbol{I}_{3\times3})\boldsymbol{\omega}\\ \dot{q}_4&=-\frac{1}{2}\boldsymbol{\varrho}^{\mathrm{T}}\boldsymbol{\omega}\end{aligned}\tag{2.18}$$

2.2.2　航天器相对姿态运动学方程

定义 $\boldsymbol{q}_{d/H}$ 和 $\boldsymbol{q}_{c/H}$ 分别为从星和主星相对于主星 LVLH 系的相对姿态四元数，即

$$\boldsymbol{q}_{d/H}=\boldsymbol{q}_d\otimes\boldsymbol{q}_H^{-1},\boldsymbol{q}_{c/H}=\boldsymbol{q}_c\otimes\boldsymbol{q}_H^{-1}\tag{2.19}$$

式中：$\boldsymbol{q}_d$，$\boldsymbol{q}_c$ 和 $\boldsymbol{q}_H$ 分别为从星、主星以及主星 LVLH 系的惯性姿态。根据相对四元数方程可得[6]

$$\dot{\boldsymbol{q}}_{d/H}=\frac{1}{2}\boldsymbol{\varXi}(\boldsymbol{q}_{d/H})\boldsymbol{\omega}_{d/H}^{d},\dot{\boldsymbol{q}}_{c/H}=\frac{1}{2}\boldsymbol{\varXi}(\boldsymbol{q}_{c/H})\boldsymbol{\omega}_{c/H}^{c}\tag{2.20}$$

式中：$\boldsymbol{\omega}_{d/H}^{d}$ 为 d 系相对于 H 系的角速度在 d 系中的分量；$\boldsymbol{\omega}_{c/H}^{c}$ 为 c 系相对于 H 系的角速度在 c 系中的分量。它们分别定义为

$$\boldsymbol{\omega}_{d/H}^{d}=\boldsymbol{\omega}_{d/I}^{d}-\boldsymbol{A}_H^d(\boldsymbol{q}_{d/H})\boldsymbol{\omega}_{H/I}^{H}\tag{2.21}$$

$$\boldsymbol{\omega}_{c/H}^{c}=\boldsymbol{\omega}_{c/I}^{c}-\boldsymbol{A}_H^c(\boldsymbol{q}_{c/H})\boldsymbol{\omega}_{H/I}^{H}\tag{2.22}$$

式中：$\boldsymbol{\omega}_{d/I}^{d}$ 和 $\boldsymbol{\omega}_{c/I}^{c}$ 分别为从星和主星的惯性角速度；$\boldsymbol{A}_H^d$ 和 $\boldsymbol{A}_H^c$ 分别为从 H 系到 d 系和 c 系的姿态阵；$\boldsymbol{\omega}_{H/I}^{H}$ 为主星 H 系的惯性角速度在 H 系中的分量，定义为

$$\boldsymbol{\omega}_{H/I}^{H}=[0\quad 0\quad \dot{\theta}]^{\mathrm{T}}\tag{2.23}$$

根据参考文献［6］，推导可得方程（2.20）的离散传播形式，即

$$\boldsymbol{q}_{d/H_{k+1}}=\overline{\boldsymbol{\Omega}}(\boldsymbol{\omega}_{d/I}^{d})\overline{\boldsymbol{\Gamma}}(\boldsymbol{\omega}_{H/I}^{H})\boldsymbol{q}_{d/H_k}\tag{2.24}$$

式中：

$$\overline{\boldsymbol{\Omega}}(\boldsymbol{\omega}_{d/I}^{d})=\begin{bmatrix}\cos\left(\frac{1}{2}\|\boldsymbol{\omega}_{d/I}^{d}\|\Delta t\right)\boldsymbol{I}_{3\times3}-[\boldsymbol{\psi}_k\times] & \boldsymbol{\psi}_k\\ -\boldsymbol{\psi}_k^{\mathrm{T}} & \cos\left(\frac{1}{2}\|\boldsymbol{\omega}_{d/I}^{d}\|\Delta t\right)\end{bmatrix}\tag{2.25}$$

$$\overline{\boldsymbol{\Gamma}}(\boldsymbol{\omega}_{H/I}^{H})=\begin{bmatrix}\cos\left(\frac{1}{2}\|\boldsymbol{\omega}_{H/I}^{H}\|\Delta t\right)\boldsymbol{I}_{3\times3}-[\boldsymbol{\zeta}_k\times] & -\boldsymbol{\zeta}_k\\ \boldsymbol{\zeta}_k^{\mathrm{T}} & \cos\left(\frac{1}{2}\|\boldsymbol{\omega}_{H/I}^{H}\|\Delta t\right)\end{bmatrix}\tag{2.26}$$

其中，

$$\begin{aligned}\boldsymbol{\psi}_k &\equiv \sin\left(\frac{1}{2}\|\boldsymbol{\omega}_{d/I}^d\|\Delta t\right)\boldsymbol{\omega}_{d/I}^d/\|\boldsymbol{\omega}_{d/I}^d\| \\ \boldsymbol{\zeta}_k &\equiv \sin\left(\frac{1}{2}\|\boldsymbol{\omega}_{H/I}^H\|\Delta t\right)\boldsymbol{\omega}_{H/I}^H/\|\boldsymbol{\omega}_{H/I}^H\|\end{aligned} \tag{2.27}$$

这里，Δt 为采样周期。

类似地，相对姿态四元数 $\boldsymbol{q}_{c/H}$的离散传播方程为

$$\boldsymbol{q}_{c/H_{k+1}} = \overline{\boldsymbol{\Omega}}(\boldsymbol{\omega}_{c/I}^c)\overline{\boldsymbol{\Gamma}}(\boldsymbol{\omega}_{H/I}^H)\boldsymbol{q}_{c/H_k} \tag{2.28}$$

这里，用 $\boldsymbol{\omega}_{c/I}^c$替代方程（2.25）中的 $\boldsymbol{\omega}_{d/I}^d$即可得到 $\overline{\boldsymbol{\Omega}}(\boldsymbol{\omega}_{c/I}^c)$。

利用 $\boldsymbol{q}_{d/H}$和 $\boldsymbol{q}_{c/H}$可计算得到从星相对于主星的相对姿态四元数 $\boldsymbol{q}_{d/c}$，即

$$\boldsymbol{q}_{d/c} = \boldsymbol{q}_{d/H} \otimes \boldsymbol{q}_{c/H}^{-1} \tag{2.29}$$

2.3 敏感器模型

2.3.1 VISNAV 观测模型

位置敏感器（Position Sensing Diode，PSD）是 20 世纪 80 年代初出现的一种新型敏感器，其信号处理简单，几何分辨率远高于传统的 CCD（Charge Coupled Device）敏感器，并具有频谱响应宽、响应速度快、能够同时测量光点（标志点）的强度和位置等优点。Texas A&M 大学 Junkins 等充分利用 PSD 技术的优点，研制出一种视觉导航系统，并相应地开发了许多相对位姿确定算法[7-10]，可以实现对运动目标相对位置和姿态的快速测量与确定。本节将着重介绍两种 VISNAV 观测模型，观测模型 1 适用于单纯依赖视觉测量方法进行相对位姿解算，观测模型 2 适用于结合航天器相对动力学设计导航滤波器。

2.3.1.1 观测模型 1

对于近距离两航天器，VISNAV 系统利用静态的几何关系可确定从星相对于主星的相对位置和姿态。图 2.2 所示为 VISNAV 系统工作示意图。摄影测量本质可由一系列共线方程描述，这里，假设敏感器坐标系与从星本体系重合，且 z 轴为光轴方向，则物平面和像平面坐标系的透视投影方程（无测量噪声）可写为

$$\alpha_i = -f\frac{A_{11}(X_i - x^c) + A_{12}(Y_i - y^c) + A_{13}(Z_i - z^c)}{A_{31}(X_i - x^c) + A_{32}(Y_i - y^c) + A_{33}(Z_i - z^c)}, i = 1, 2, \cdots, N \tag{2.30}$$

$$\gamma_i = -f\frac{A_{21}(X_i - x^c) + A_{22}(Y_i - y^c) + A_{23}(Z_i - z^c)}{A_{31}(X_i - x^c) + A_{32}(Y_i - y^c) + A_{33}(Z_i - z^c)}, i = 1, 2, \cdots, N \tag{2.31}$$

式中：N 为总观测量个数；(α_i, γ_i) 为第 i 个标志点的像平面坐标测量值；(X_i, Y_i, Z_i) 为已知的第 i 个标志点在主星体系中的位置分量；(x^c, y^c, z^c) 为待求的相对位置矢量在主星体系中的分量；f 为已知的相机焦距；A_{jk} 为待求的相对姿态阵 $\boldsymbol{A}_c^d(\boldsymbol{q}_{d/c})$ 的元素。

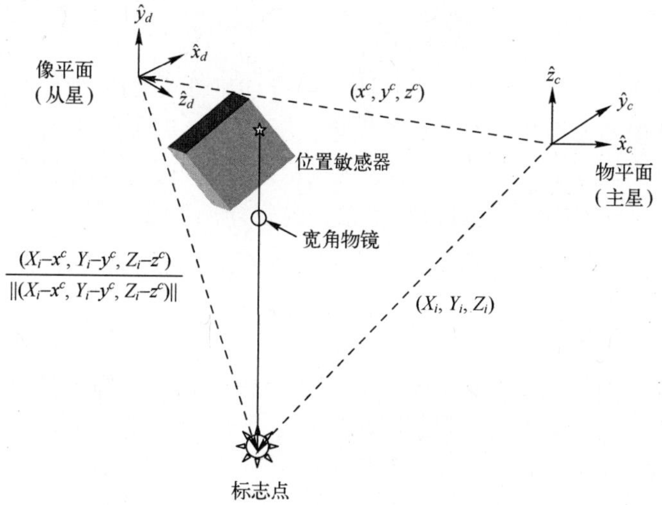

图 2.2　VISNAV 系统（观测模型 1）

VISNAV 系统作为一类视线测量敏感器，上述测量值通常可以重构为单位矢量形式，即第 i 个视线观测量可写为

$$\boldsymbol{b}_i = \boldsymbol{A}_c^d(\boldsymbol{q}_{d/c})\boldsymbol{r}_i \tag{2.32}$$

其中，

$$\boldsymbol{b}_i \equiv \frac{1}{\sqrt{f^2+\alpha_i^2+\gamma_i^2}}\begin{bmatrix} -\alpha_i \\ -\gamma_i \\ f \end{bmatrix} \tag{2.33}$$

$$\boldsymbol{r}_i \equiv \frac{1}{\sqrt{(X_i-x^c)^2+(Y_i-y^c)^2+(Z_i-z^c)^2}}\begin{bmatrix} (X_i-x^c) \\ (Y_i-y^c) \\ (Z_i-z^c) \end{bmatrix} \tag{2.34}$$

另外，若已知视线观测矢量 $\boldsymbol{b}_i$，则由式（2.33）可得标志点的像平面坐标为

$$\alpha_i = -f \cdot \frac{b_{ix}}{b_{iz}}, \gamma_i = -f \cdot \frac{b_{iy}}{b_{iz}} \tag{2.35}$$

式中：b_{ix}，b_{iy} 和 b_{iz} 为观测矢量 $\boldsymbol{b}_i$ 的三分量。

考虑敏感器测量噪声后，式（2.32）可进一步改写为

$$\tilde{\boldsymbol{b}}_i=\boldsymbol{A}_c^d(\boldsymbol{q}_{d/c})\boldsymbol{r}_i+\boldsymbol{v}_i,\quad \boldsymbol{v}_i^{\mathrm{T}}\boldsymbol{A}_c^d\boldsymbol{r}_i=0 \tag{2.36}$$

式中：$\tilde{\boldsymbol{b}}_i$为在 d 系中第 i 个标志点的单位观测矢量；$\boldsymbol{v}_i$ 为敏感器测量噪声，近似为高斯噪声，满足

$$\boldsymbol{E}\{\boldsymbol{v}_i\}=\boldsymbol{0} \tag{2.37}$$

$$\boldsymbol{E}\{\boldsymbol{v}_i\boldsymbol{v}_i^{\mathrm{T}}\}=\sigma_i^2[\boldsymbol{I}_{3\times3}-(\boldsymbol{A}_c^d\boldsymbol{r}_i)(\boldsymbol{A}_c^d\boldsymbol{r}_i)^{\mathrm{T}}] \tag{2.38}$$

式中：σ_i为测量噪声标准差。这就是著名的 QUEST 观测模型，对于小视场敏感器而言该误差模型非常准确，其优势在于观测方差阵在 EKF 公式中可以等效地用非奇异阵 $\sigma_i^2\boldsymbol{I}_{3\times3}$代替。针对大视场敏感器情形，Cheng 等利用一阶泰勒展开近似扩展了 QUEST 测量模型，相关内容详见参考文献［11］，这里不再赘述。

2.3.1.2 观测模型 2

结合航天器相对动力学模型和 VISNAV 系统，可设计单目视觉导航滤波器。值得注意的是，由于标志点定义在主星本体系，而相对轨道坐标定义在主星 LVLH 系，这里利用了相对四元数 $\boldsymbol{q}_{d/H}$和 $\boldsymbol{q}_{c/H}$构造视线观测量，推导了第二种 VISNAV 观测模型。不同于参考文献［12］中的观测模型，该模型并不做主星本体系与其 LVLH 系重合这一假设。图 2.3 所示为相应的工作示意图。

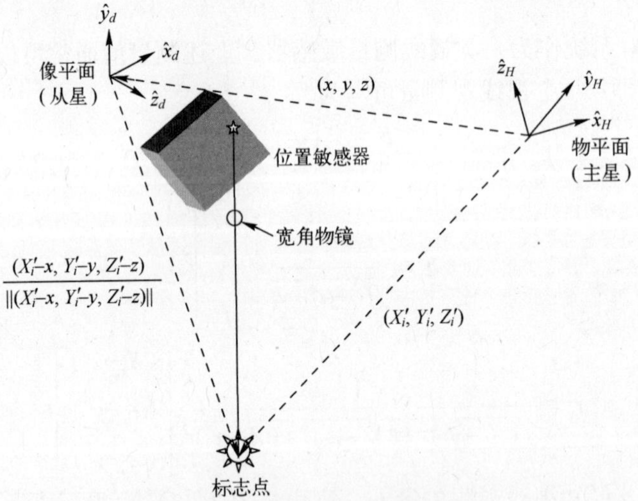

图 2.3 VISNAV 系统（观测模型 2）

因此，对于观测模型 2，第 i 个视线观测量可写为

$$\tilde{\boldsymbol{b}}_i=\boldsymbol{A}_H^d(\boldsymbol{q}_{d/H})\boldsymbol{r}_i+\boldsymbol{v}_i,\quad \boldsymbol{v}_i^{\mathrm{T}}\boldsymbol{A}_H^d\boldsymbol{r}_i=0 \tag{2.39}$$

其中，

$$\boldsymbol{r}_i \equiv \frac{\boldsymbol{\chi}'_i-\boldsymbol{\rho}}{\|\boldsymbol{\chi}'_i-\boldsymbol{\rho}\|}=\frac{1}{\sqrt{(X'_i-x)^2+(Y'_i-y)^2+(Z'_i-z)^2}}\begin{bmatrix}(X'_i-x)\\(Y'_i-y)\\(Z'_i-z)\end{bmatrix} \tag{2.40}$$

式中：$\widetilde{\boldsymbol{b}}_i$为在 d 系中第 i 个标志点的单位观测矢量；$\boldsymbol{v}_i$为敏感器测量噪声；$\boldsymbol{\rho}\equiv[x,y,z]^{\mathrm{T}}$为式（2.1）中的相对轨道坐标；$\boldsymbol{\chi}'_i\equiv[X'_i,Y'_i,Z'_i]^{\mathrm{T}}$为第 i 个标志点的位置矢量在 H 系中的描述，即

$$\boldsymbol{\chi}'_i=[\boldsymbol{A}_H^c(\boldsymbol{q}_{c/H})]^{\mathrm{T}}\boldsymbol{\chi}_i \tag{2.41}$$

其中，$\boldsymbol{\chi}_i\equiv[X_i,Y_i,Z_i]^{\mathrm{T}}$为已知第 i 个标志点的位置矢量在主星体系中的分量。

2.3.2　陀螺测量模型

一般来说，陀螺测量的是航天器的惯性姿态角速度在体系中的分量，通常采用模型为[13]

$$\begin{aligned}\widetilde{\boldsymbol{\omega}}&=\boldsymbol{\omega}+\boldsymbol{\beta}+\boldsymbol{\eta}_v\\ \dot{\boldsymbol{\beta}}&=\boldsymbol{\eta}_u\end{aligned} \tag{2.42}$$

式中：$\widetilde{\boldsymbol{\omega}}$为陀螺的测量输出；$\boldsymbol{\omega}$ 为惯性角速度真值；$\boldsymbol{\beta}$ 为漂移量；$\boldsymbol{\eta}_v$和 $\boldsymbol{\eta}_u$为独立的零均值高斯白噪声过程，满足

$$\begin{aligned}\boldsymbol{E}\{\boldsymbol{\eta}_v(t)\boldsymbol{\eta}_v^{\mathrm{T}}(\tau)\}&=\sigma_v^2\delta(t-\tau)\boldsymbol{I}_{3\times3}\\ \boldsymbol{E}\{\boldsymbol{\eta}_u(t)\boldsymbol{\eta}_u^{\mathrm{T}}(\tau)\}&=\sigma_u^2\delta(t-\tau)\boldsymbol{I}_{3\times3}\end{aligned} \tag{2.43}$$

式中：σ_v和 σ_u分别为随机漂移噪声 $\boldsymbol{\eta}_v$和漂移斜率噪声 $\boldsymbol{\eta}_u$的标准差，更为一般地，考虑了刻度因子和失准角误差的陀螺模型可参见参考文献［14］。

在数值仿真过程中，离散形式的陀螺测量值可由递推公式产生[15]，即

$$\widetilde{\boldsymbol{\omega}}_{k+1}=\boldsymbol{\omega}_{k+1}+\frac{1}{2}(\boldsymbol{\beta}_{k+1}+\boldsymbol{\beta}_k)+\left(\frac{\sigma_v^2}{\Delta t}+\frac{1}{12}\sigma_u^2\Delta t\right)^{1/2}\boldsymbol{N}_{vk} \tag{2.44}$$

$$\boldsymbol{\beta}_{k+1}=\boldsymbol{\beta}_k+\sigma_u\Delta t^{1/2}\boldsymbol{N}_{uk} \tag{2.45}$$

式中：下标 k 表示第 k 个采样时刻；Δt 为陀螺采样周期；$\boldsymbol{N}_{vk}$和 $\boldsymbol{N}_{uk}$为独立的零均值高斯白噪声过程，其协方差阵均为单位阵。

2.4　执行机构模型

2.4.1　喷气推力器

喷气推力器系统通常具有反应快、精度高、机动能力强，且能适应各种环境等优点，而且推力的大小、喷管的安装位置都允许在较大范围内选择，

因此，小推力器系统常被采用且能够给控制系统的设计带来方便。不过，小推力器系统需要消耗工质，不允许长期工作，因此在应用中又受到限制。根据小推力器系统的特点，许多航天器的喷气推力器系统是兼用于质心控制与姿态控制的，尤其是在空间交会、对接以及返回等过程中，此时控制与制导往往是同时进行的。统一考虑六自由度的控制，可组成更优的控制律，以提高性能和节省工质。

由于推力器产生的推力是属于开关型的控制量，因而采用相平面开关曲线方法设计的控制律获得广泛应用。近年来，长寿命、高可靠性、重复性能好的推力器系统投入使用，其极限的开关次数高达数百万次。因此，若采用脉宽脉频（Pulse-Width Pulse-Frequency，PWPF）调制方式，把开关型控制量调制成连续型控制量，则控制律较为简单且不需要复杂的计算。

PWPF 调节器，通过在继电特性环节前面串联一个一阶惯性环节，将脉冲的频率和宽度进行自动调制，从而把开关型控制量等效为连续型控制量。图 2.4 所示为 PWPF 调节器原理。

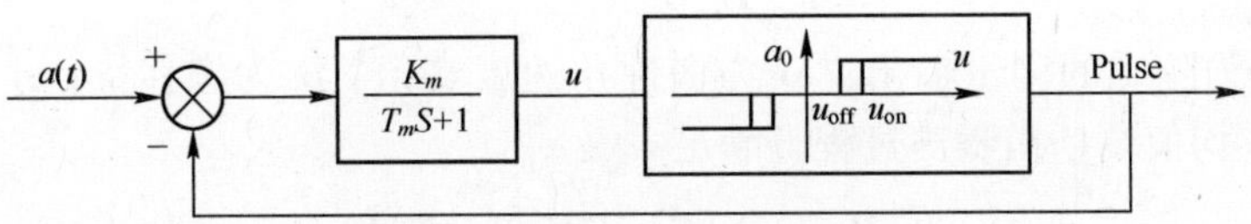

图 2.4　PWPF 调节器原理图

图 2.4 中，K_m 和 T_m 分别为一阶惯性环节的放大系数和时间常数；u_{on} 和 u_{off} 分别为继电器开和关的门限值；a_0 为继电器脉冲幅值；a 为连续型输入量。由于调节器能够把开关量调制为连续量，才使得变推力控制律的研究具有实际意义，更多相关原理详见参考文献［16］。

2.4.2　控制力矩陀螺

当今常用的航天器姿态执行机构主要有反作用飞轮、推力器、单框架控制力矩陀螺（Single-gimbal Control Moment Gyro，SGCMG）和双框架控制力矩陀螺（Double-gimbal Control Moment Gyro，DGCMG）等。与反作用飞轮一样，SGCMG 也属于一种基于角动量交换的姿态执行机构。反作用飞轮依靠的是控制飞轮转子转速增减的方式来产生控制力矩，而 SGCMG 依靠的是框架带动恒速飞轮转动使得飞轮的角动量方向发生改变来产生进动控制力矩。与反作用飞轮相比，SGCMG 具有力矩放大能力，可以以更小的质量和体积输出更大的控制力矩，实现航天器的快速姿态机动。与常用的推力器相比，SGCMG 具有不消耗燃料、不污染光学设备的优点；与 DGCMG 相比，SGCMG 具有机械结构简单和可靠性高的优点。因此，SGCMG 成了当今最具有吸引力的航天器姿态执行机构。

2.4.2.1 SGCMG 系统的数学模型

根据欧拉定理，采用控制力矩陀螺（Control Moment Gyro，CMG）作为执行机构的刚体航天器的姿态动力学方程可描述为[17]

$$\boldsymbol{J}\dot{\boldsymbol{\omega}}+[\boldsymbol{\omega}\times]\boldsymbol{J}\boldsymbol{\omega}=\boldsymbol{T}+\boldsymbol{d} \tag{2.46}$$

$$\dot{\boldsymbol{h}}+\boldsymbol{\omega}\times\boldsymbol{h}=-\boldsymbol{T} \tag{2.47}$$

式中：$\boldsymbol{J}$ 为航天器的惯量张量矩阵；$\boldsymbol{h}$ 为总 CMG 角动量矢量；$\dot{\boldsymbol{h}}$ 为角动量变化率；$\boldsymbol{T}$ 为由 CMG 产生的内部控制力矩；$\boldsymbol{d}$ 为作用在航天器上的外部干扰力矩矢量，如重力梯度力矩、气动力矩和太阳光压力矩等。

为了实现航天器三轴姿态控制，姿态控制系统至少需要三个 SGCMG 单元。金字塔安装构型采用 4 个 SGCMG 对称安装，具有最小的冗余度，安装构型如图 2.5 所示。不失一般性，假设每个 SGCMG 单元的角动量均为 1N · m · s，那么金字塔构型 SGCMG 系统的角动量在本体系的三轴分量为

$$\begin{aligned}\boldsymbol{h}(\boldsymbol{\delta})&=\sum_{i=1}^{4}\boldsymbol{H}_i(\boldsymbol{\delta}_i)\\&=\begin{bmatrix}-c\beta\sin\delta_1\\ \cos\delta_1\\ s\beta\sin\delta_1\end{bmatrix}+\begin{bmatrix}-\cos\delta_2\\ -c\beta\sin\delta_2\\ s\beta\sin\delta_2\end{bmatrix}+\begin{bmatrix}c\beta\sin\delta_3\\ -\cos\delta_3\\ s\beta\sin\delta_3\end{bmatrix}+\begin{bmatrix}\cos\delta_4\\ c\beta\sin\delta_4\\ s\beta\sin\delta_4\end{bmatrix}\end{aligned} \tag{2.48}$$

式中：$\boldsymbol{H}_i$ 为第 i 个 SGCMG 在本体系中描述的角动量矢量；β 为安装倾角，$c\beta\equiv\cos\beta$，$s\beta\equiv\sin\beta$；$\delta_i(i=1,2,3,4)$ 为第 i 个框架角。

金字塔构型 SGCMG 系统总角动量 $\boldsymbol{h}$ 的时间导数为

$$\dot{\boldsymbol{h}}=\sum_{i=1}^{4}\dot{\boldsymbol{H}}_i=\sum_{i=1}^{4}\boldsymbol{c}_i(\delta_i)\dot{\delta}_i=\boldsymbol{C}(\boldsymbol{\delta})\dot{\boldsymbol{\delta}} \tag{2.49}$$

式中：$\dot{\delta}_i$ 为第 i 个框架角速率；$\boldsymbol{\delta}=[\delta_1\quad\delta_2\quad\delta_3\quad\delta_4]^{\mathrm{T}}$ 为框架角矢量；$\boldsymbol{c}_i$ 为雅可比矩阵 $\boldsymbol{C}$ 的第 i 列，矩阵 $\boldsymbol{C}$ 为

$$\boldsymbol{C}(\boldsymbol{\delta})=\begin{bmatrix}-c\beta\cos\delta_1 & \sin\delta_2 & c\beta\cos\delta_3 & -\sin\delta_4\\ -\sin\delta_1 & -c\beta\cos\delta_2 & \sin\delta_3 & c\beta\cos\delta_4\\ s\beta\cos\delta_1 & s\beta\cos\delta_2 & s\beta\cos\delta_3 & s\beta\cos\delta_4\end{bmatrix} \tag{2.50}$$

奇异问题是指 SGCMG 系统在某种框架角矢量 $\boldsymbol{\delta}$ 配置情况下，每个 SGCMG 所能提供的力矩均正交于期望的控制力矩 $\boldsymbol{\tau}_d$，失去在该方向输出力矩的能力。定义表示奇异程度大小的奇异度量为

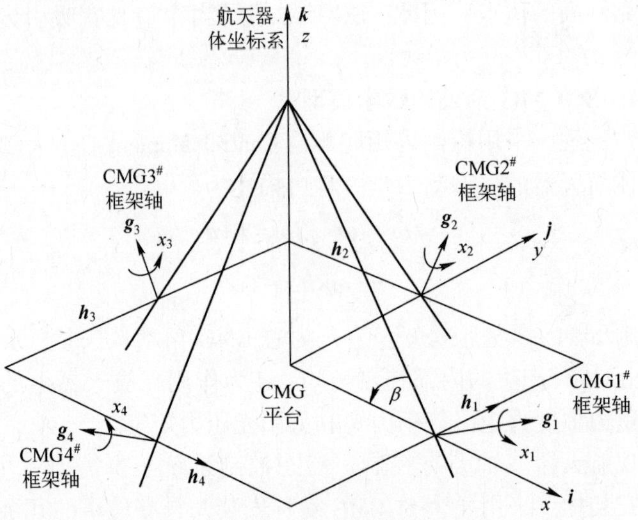

图 2.5　金字塔构型 SGCMG 系统

$$D=\det(\boldsymbol{C}\boldsymbol{C}^{\mathrm{T}}) \tag{2.51}$$

$D=0$ 表示 SGCMG 系统完全陷入奇异状态，D 越大表明 SGCMG 系统距奇异状态越远。金字塔构型的 SGCMG 系统角动量包络内含有大量奇异点，导致整个执行机构非常容易陷入奇异状态而无法正常输出力矩，因此需要设计适当的操纵律来回避奇异状态。

2.4.2.2　SGCMG 操纵律设计

针对控制力矩 $\boldsymbol{T}$，总 CMG 角动量速率指令或者力矩指令为

$$\dot{\boldsymbol{h}}\equiv\boldsymbol{\tau}_d=-\boldsymbol{T}-\boldsymbol{\omega}\times\boldsymbol{h} \tag{2.52}$$

SGCMG 操纵律的任务就是要根据力矩指令 $\boldsymbol{\tau}_d$ 和当前 SGCMG 系统的框架角状态，在满足硬件输出能力的条件下合理分配各个 SGCMG 单元的框架角速度，使得整个 SGCMG 系统在精确输出期望力矩的同时有效回避奇异状态。因此，评价一个操纵律的性能主要看两个方面：①力矩输出误差的大小；②回避和逃离奇异状态的能力。下面介绍常见的两种操纵律，它们是标准伪逆操纵律和奇异鲁棒操纵律[17-18]。

1. 标准伪逆操纵律

根据力矩指令 $\boldsymbol{\tau}_d$ 和框架角速度 $\dot{\boldsymbol{\delta}}$ 的关系可知，当 $\mathrm{rank}(\boldsymbol{C})=3$ 时，对应的给定解不唯一，从控制量范数角度出发，设计函数为

$$f=\frac{1}{2}\dot{\boldsymbol{\delta}}^{\mathrm{T}}\dot{\boldsymbol{\delta}} \tag{2.53}$$

求解式（2.53）约束下的f函数极值，可得到系统的伪逆解，即$\dot{\boldsymbol{\delta}}$的最小范数解。此时，方程的求解转换为等式约束下的极值问题，引入拉格朗日乘子$\boldsymbol{\lambda}$，选定指标函数为

$$J=\frac{1}{2}\dot{\boldsymbol{\delta}}^{\mathrm{T}}\dot{\boldsymbol{\delta}}+\boldsymbol{\lambda}^{\mathrm{T}}(\boldsymbol{\tau}_d-\boldsymbol{C}\dot{\boldsymbol{\delta}}) \tag{2.54}$$

将指标函数对$\dot{\boldsymbol{\delta}}$和$\boldsymbol{\lambda}$求偏导后取零，得到系统的最小范数解为

$$\dot{\boldsymbol{\delta}}=\boldsymbol{C}^{+}\boldsymbol{\tau}_d,\quad \boldsymbol{C}^{+}=\boldsymbol{C}^{\mathrm{T}}(\boldsymbol{C}\boldsymbol{C}^{\mathrm{T}})^{-1} \tag{2.55}$$

这样就可以根据给定力矩指令得到控制量消耗最小值对应的唯一解，但在航天器姿态控制过程中，指令力矩的变化趋势是不可预知的，因而难以预估SGCMG系统的角动量及框架构型的变化趋向。式（2.55）需要保证矩阵$\boldsymbol{C}\boldsymbol{C}^{\mathrm{T}}$可逆，而当系统到达或邻近奇异状态时（$\mathrm{rank}(\boldsymbol{C})<3$），伪逆解对应即为无解或得到的指令角速度极大，远远超出现有执行机构的能力，因此这种操纵律很难满足实际应用的需要。

2. 奇异鲁棒操纵律

当SGCMG系统处于奇异状态时，$\boldsymbol{C}\boldsymbol{C}^{\mathrm{T}}$不可逆，所以方程无法计算伪逆解。此时，SGCMG系统仍可在某一平面内输出力矩，所以可以在操纵律中引入小量的干扰误差来避开即将遇到的奇异状态，但这会牺牲一定的操纵精度，因此必须在避免奇异性与操纵精度之间做出权衡。

如果$\mathrm{rank}(\boldsymbol{C})<3$，则有$\det(\boldsymbol{C}\boldsymbol{C}^{\mathrm{T}})=0$，那么伪逆解$\boldsymbol{C}^{+}$不存在。为了确定此状态下的SGCMG框架角速度，考虑最小二次型问题为

$$\min_{\dot{\boldsymbol{\delta}}}\{(\boldsymbol{C}\dot{\boldsymbol{\delta}}-\boldsymbol{\tau}_d)^{\mathrm{T}}\boldsymbol{P}(\boldsymbol{C}\dot{\boldsymbol{\delta}}-\boldsymbol{\tau}_d)+\dot{\boldsymbol{\delta}}^{\mathrm{T}}\boldsymbol{Q}\dot{\boldsymbol{\delta}}\} \tag{2.56}$$

式中：$\boldsymbol{P}$和$\boldsymbol{Q}$为正定对称加权矩阵。求解式（2.56）极值即可得到奇异鲁棒伪逆解

$$\dot{\boldsymbol{\delta}}=\boldsymbol{C}^{\#}\boldsymbol{\tau}_d,\quad \boldsymbol{C}^{\#}=[\boldsymbol{C}^{\mathrm{T}}\boldsymbol{P}\boldsymbol{C}+\boldsymbol{Q}]^{-1}\boldsymbol{C}^{\mathrm{T}}\boldsymbol{P} \tag{2.57}$$

若$\boldsymbol{Q}=0$，则奇异鲁棒伪逆解退化为最小二乘解

$$\boldsymbol{C}^{\#}=[\boldsymbol{C}^{\mathrm{T}}\boldsymbol{P}\boldsymbol{C}]^{-1}\boldsymbol{C}^{\mathrm{T}}\boldsymbol{P} \tag{2.58}$$

若按如下方式选取$\boldsymbol{P}$和$\boldsymbol{Q}$，$\boldsymbol{P}=\boldsymbol{I}_{3\times3}$，$\boldsymbol{Q}=\lambda\boldsymbol{I}_{4\times4}$，则奇异鲁棒伪逆解的形式为

$$\dot{\boldsymbol{\delta}}=\boldsymbol{C}^{\#}\boldsymbol{\tau}_d \tag{2.59}$$

其中，

$$\boldsymbol{C}^{\#}=[\boldsymbol{C}^{\mathrm{T}}\boldsymbol{C}+\lambda\boldsymbol{I}_{4\times4}]^{-1}\boldsymbol{C}^{\mathrm{T}}\equiv\boldsymbol{C}^{\mathrm{T}}[\boldsymbol{C}\boldsymbol{C}^{\mathrm{T}}+\lambda\boldsymbol{I}_{3\times3}]^{-1} \tag{2.60}$$

式中：λ 为待定的奇异逃离参数。

当系统远离奇异时，λ 可以取一个小量，以减少力矩输出误差；当系统接近奇异时，λ 应迅速增大，以驱使系统迅速逃离奇异。因此，可将参数选取为

$$\lambda=\lambda_0\exp[-\mu\det(\boldsymbol{C}\boldsymbol{C}^{\mathrm{T}})] \tag{2.61}$$

式中：λ_0 和 μ 为待定的参数。

另外，针对式（2.60），一种广义的奇异鲁棒操纵律形式为

$$\boldsymbol{C}^{\#}=[\boldsymbol{C}^{\mathrm{T}}\boldsymbol{P}\boldsymbol{C}+\lambda\boldsymbol{I}_{4\times4}]^{-1}\boldsymbol{C}^{\mathrm{T}}\boldsymbol{P}=\boldsymbol{C}^{\mathrm{T}}[\boldsymbol{C}\boldsymbol{C}^{\mathrm{T}}+\lambda\boldsymbol{E}]^{-1} \tag{2.62}$$

$$\boldsymbol{P}^{-1}\equiv\boldsymbol{E}=\begin{bmatrix}1 & \varepsilon_3 & \varepsilon_2\\ \varepsilon_3 & 1 & \varepsilon_1\\ \varepsilon_2 & \varepsilon_1 & 1\end{bmatrix}>0 \tag{2.63}$$

式中：参数 λ 和 ε_i 应该合理选择，使得对于任何非零力矩指令 $\boldsymbol{\tau}_d$，均能保证 $\boldsymbol{C}^{\#}\boldsymbol{\tau}_d\neq0$。通常，$\varepsilon_i(i=1,2,3)$ 可选择为周期正弦函数：

$$\varepsilon_i=\varepsilon_0\sin(\omega t+\phi_i) \tag{2.64}$$

其中：振幅 ε_0、调制频率 ω 和相位 ϕ_i 为待定参数。

奇异鲁棒操纵律在伪逆操纵律的基础上添加了 $\lambda\boldsymbol{E}$ 项，这样可以以带来一定的力矩误差为代价保证在奇异状态下$[\boldsymbol{C}\boldsymbol{C}^{\mathrm{T}}+\lambda\boldsymbol{E}]^{-1}$的逆存在，此时奇异鲁棒操纵律能够继续输出框架角速度来逃离奇异。但是，奇异鲁棒操纵律存在“框架角锁定现象”，即在某些奇异状态上，当期望力矩矢量与奇异方向一致时，奇异鲁棒操纵律输出的框架角速度始终为零。此时，整个系统将始终处于奇异状态，无法逃离，这是奇异鲁棒操纵律的最大缺点。为了克服这个问题，Wie 进一步改进了加权矩阵 $\boldsymbol{Q}$，提出了非奇异鲁棒操纵律，相关内容详见参考文献［19］。

参 考 文 献

［1］SCHAUB H, JUNKINS J L. Analytical mechanices of aerospace systems［M］. New York：AIAA Inc, 2003.

［2］SHUSTER M D. A survey of attitude representations［J］. Journal of the Astronautical Sciences, 1993, 41（4）：439-517.

［3］WERTZ J R. Spacecraft attitude determination and control［M］. Dordrecht：Kluwer Academic Publishers, 1978.

［4］LEFFERTS E J, MARKLEY F L, SHUSTER M D. Kalman filtering for spacecraft attitude estimation [J]. Journal of Guidance, Control, and Dynamics, 1982, 5 (5): 417-429.

[5] HAMILTON W R. Elements of quaternions [M]. London: Longmans, Green and Co., 1866.

[6] MAYO R A. Relative quaternion state transition relation [J]. Journal of Guidance, Control, and Dynamics, 1979, 2 (1): 44-48.

[7] JUNKINS J L, HUGHES D C, WAZNI K P, et al. Vision-based navigation for rendezvous and docking and proximity operations [C]//Proceedings of the annual AAS Rocky Mountain Guidance and Control Conference, Breckenridge. USA: AIAA Inc, 1999: 203-220.

[8] CRASSIDIS J L, ALONSO R, JUNKINS J L. Optimal attitude and position determination from line-of-sight measurements [J]. Journal of the Astronautical Sciences, 2000, 48: 391-408.

[9] MORTARI D, ROJAS J M, JUNKINS J L. Attitude and position estimation from vector observations [C]//The 14th AAS/AIAA Space Flight Mechanics Meeting, Feb 8-12, 2004. USA: AIAA Inc, 2004: 1-18.

[10] HAAS B M. Sensitivity study: the effects of beacon location errors on a vehicle's position and attitude estimation for a vision-based navigation system [D]. New York: State University of New York at Buffalo, 2005.

[11] CHENG Y, CRASSIDIS J L, MARKLEY F L. Attitude estimation for large field-of-view sensors [J]. Journal of the Astronautical Sciences, 2006, 54: 433-448.

[12] KIM S G, CRASSIDIS J L, CHENG Y, et al. Kalman filtering for relative spacecraft attitude and position estimation [J]. Journal of Guidance, Control, and Dynamics, 2007, 30 (1): 133-143.

[13] FARRENKOPF R L. Analytic steady-state accuracy solutions for two common spacecraft attitude estimators [J]. Journal of Guidance and Control, 1978, 1 (4): 282-284.

[14] PITTELKAU M E. Kalman filtering for spacecraft system alignment calibration [J]. Journal of Guidance, Control, and Dynamics, 2001, 24 (6): 1187-1195.

[15] CRASSIDIS J L. Sigma-point Kalman filtering for integrated gps and inertial navigation [C]// AIAA Guidance, Navigation, and Control Conference, San Francisco. USA: AIAA Inc, 2005: 1981-2004.

[16] 黄圳圭. 航天器姿态动力学 [M]. 长沙: 国防科技大学出版社, 1997.

[17] WIE B. Space vehicle dynamics and control [M]. Reston: AIAA Inc, 1998.

[18] WIE B, HEIBERG C, BAILEY D. Singularity robust steering logic for redundant single-gimbal control moment gyros [J]. Journal of Guidance, Control, and Dynamics, 2001, 24 (5): 865-872.

[19] WIE B. Singularity escape/avoidance steering logic for control moment gyro systems [J]. Journal of Guidance, Control, and Dynamics, 2005, 28 (5): 948-956.

第3章　基于非线性最小二乘法的航天器相对位姿确定

基于前述的 VISNAV 观测模型，本章将深入研究基于标志点静态几何关系的航天器相对位姿确定方法，采用非线性最小二乘法求解航天器相对位置和姿态，为后续设计的导航滤波器提供初值。考虑到四元数需要遵循单位范数约束条件，本章将选取三分量的修正罗德里格参数（Modified Rodrigues Parameters，MRPs）[1]作为姿态描述参数，更有利于非线性最小二乘法进行相对位姿解算。综合考虑标志点位置误差噪声和 VISNAV 系统测量噪声两方面的影响，推导观测误差噪声阵并将其用于非线性最小二乘算法中。

3.1　基于 MRPs 的 VISNAV 观测模型

3.1.1　修正罗德里格参数

修正罗德里格参数是一组三分量姿态描述参数，可由四元数转换得到[2]

$$\boldsymbol{p} \equiv \boldsymbol{\varrho}/(1+q_4) = \boldsymbol{e}\tan(\boldsymbol{\phi}/4) \tag{3.1}$$

式中：$\boldsymbol{p}=[p_1 \quad p_2 \quad p_3]^{\mathrm{T}}$；$\boldsymbol{e}$ 为单位欧拉旋转轴；$\boldsymbol{\phi}$ 为旋转角。分析可知，当 $\|\boldsymbol{\phi}\|\to 2\pi$，使得 $\|\boldsymbol{p}\|\to\infty$，MRPs 出现奇异。由于四元数在旋转群中满足 2∶1 的匹配关系，即四元数 $\boldsymbol{q}$ 和 $-\boldsymbol{q}$ 描述同一个旋转关系，那么根据式（3.1）同样可以定义影子 MRPs $\boldsymbol{p}^s=-\boldsymbol{p}/|\boldsymbol{p}|^2$。显然，利用 $\boldsymbol{p}$ 和 $\boldsymbol{p}^s$ 之间的切换，MRPs 可实现全局非奇异姿态描述。

相应地，四元数也可以由 MRPs 转换得到，即

$$\boldsymbol{q}=\pm\frac{1}{1+|\boldsymbol{p}|^2}\begin{bmatrix} 2\boldsymbol{p} \\ 1-|\boldsymbol{p}|^2 \end{bmatrix} \tag{3.2}$$

姿态矩阵可用 MRPs 描述为

$$\boldsymbol{A}(\boldsymbol{p})=\boldsymbol{I}_{3\times3}-4\frac{(1-\boldsymbol{p}^{\mathrm{T}}\boldsymbol{p})}{(1+\boldsymbol{p}^{\mathrm{T}}\boldsymbol{p})^2}[\boldsymbol{p}\times]+\frac{8}{(1+\boldsymbol{p}^{\mathrm{T}}\boldsymbol{p})^2}[\boldsymbol{p}\times]^2 \tag{3.3}$$

进一步，由 MRPs 表示的姿态运动学方程为[2]

$$\dot{\boldsymbol{p}}=\boldsymbol{G}(\boldsymbol{p})\boldsymbol{\omega} \tag{3.4}$$

其中,

$$G(p)=\frac{1}{4}\{(1-p^{\mathrm{T}}p)I_{3\times3}+2[p\times]+2pp^{\mathrm{T}}\} \tag{3.5}$$

3.1.2 VISNAV 观测模型 1（含噪声）

对于 2.3.1 节中描述的 VISNAV 观测模型 1，本节选取三分量的 MRPs 用于描述姿态阵，则 VISNAV 模型的测量方程为[3]

$$\alpha_i=-f\frac{A_{11}(X_i-x^c)+A_{12}(Y_i-y^c)+A_{13}(Z_i-z^c)}{A_{31}(X_i-x^c)+A_{32}(Y_i-y^c)+A_{33}(Z_i-z^c)},i=1,2,\cdots,N \tag{3.6}$$

$$\gamma_i=-f\frac{A_{21}(X_i-x^c)+A_{22}(Y_i-y^c)+A_{23}(Z_i-z^c)}{A_{31}(X_i-x^c)+A_{32}(Y_i-y^c)+A_{33}(Z_i-z^c)},i=1,2,\cdots,N \tag{3.7}$$

式中：N 为标志点总数；(α_i,γ_i) 为第 i 个标志点的像平面坐标测量值；(X_i,Y_i,Z_i) 为已知的第 i 个标志点在主星体系中的位置分量；(x^c,y^c,z^c) 为待求的相对位置矢量在主星体系中的分量；f 为已知的相机焦距；A_{jk} 为待求的相对姿态阵 $A_c^d(p_{d/c})$ 的元素。

进一步，考虑到安装标志点时，航天器上柔性部件的振动会造成标志点的实际安装位置与标称值之间存在偏差。假设该误差是零均值高斯噪声，则标志点位置模型为

$$\tilde{\boldsymbol{\mathcal{X}}}_i=\boldsymbol{\mathcal{X}}_i+\boldsymbol{v}_{1i} \tag{3.8}$$

式中：$\tilde{\boldsymbol{\mathcal{X}}}_i\equiv[\tilde{X}_i,\tilde{Y}_i,\tilde{Z}_i]^{\mathrm{T}}$ 和 $\boldsymbol{\mathcal{X}}_i\equiv[X_i,Y_i,Z_i]^{\mathrm{T}}$ 分别为第 i 个标志点的实际安装位置和已知的标称安装位置；$\boldsymbol{v}_{1i}$ 为第 i 个标志点的位置误差噪声，标准差为 σ_1。

值得注意的是，对于标志点的像平面坐标测量值 α_i 和 γ_i，式（3.6）和式（3.7）中并没有包含测量噪声。因此，综合考虑可得含噪声的 VISNAV 观测模型为

$$\tilde{y}=\tilde{c}=f(x,v_1,v_2)=f(x)+w_1(x,v_1)+w_2(v_2) \tag{3.9}$$

式中：

$$\tilde{c}=\begin{bmatrix}\tilde{c}_1\\ \tilde{c}_2\\ \vdots\\ \tilde{c}_N\end{bmatrix}\equiv\begin{bmatrix}\tilde{\alpha}_1\\ \tilde{\gamma}_1\\ \tilde{\alpha}_2\\ \tilde{\gamma}_2\\ \vdots\\ \tilde{\alpha}_N\\ \tilde{\gamma}_N\end{bmatrix}_{2N\times1},\quad x=\begin{bmatrix}s^c\\ p_{d/c}\end{bmatrix}\equiv\begin{bmatrix}x^c\\ y^c\\ z^c\\ p_1\\ p_2\\ p_3\end{bmatrix}_{6\times1} \tag{3.10}$$

$\boldsymbol{f}(\cdot)$为透视投影方程组；$\boldsymbol{w}_1(\cdot)$和$\boldsymbol{w}_2(\cdot)$为过程噪声函数（零均值高斯噪声）；$\boldsymbol{v}_1$为标志点位置误差噪声矢量，标准差是σ_1；$\boldsymbol{v}_2$为 VISNAV 系统的测量噪声矢量，标准差是σ_2。

在数值仿真中，标志点的像平面坐标真值$\boldsymbol{c}_i$是将状态量真值$\boldsymbol{x}$和实际的标志点位置矢量$\widetilde{\boldsymbol{\mathcal{X}}}_i$代入透视投影方程组得到的，而测量值$\widetilde{\boldsymbol{c}}_i$是在此基础上加上 VISNAV 系统测量噪声得到的。因此，第i个标志点的像平面坐标测量方程可写为

$$\widetilde{\boldsymbol{c}}_i=\begin{bmatrix}\widetilde{\alpha}_i\\ \widetilde{\gamma}_i\end{bmatrix}=\boldsymbol{f}(\boldsymbol{x},\widetilde{\boldsymbol{\mathcal{X}}}_i(\boldsymbol{\mathcal{X}}_i,\boldsymbol{v}_{1i}))+\boldsymbol{v}_{2i},\quad i=1,2,\cdots,N \tag{3.11}$$

式中：$\boldsymbol{v}_{2i}$为第i个标志点的像平面坐标测量噪声。

3.2　基于非线性最小二乘法的相对位姿确定方法

3.2.1　非线性最小二乘算法

一般来说，在估计过程中会涉及真值、测量值和估计值三种变量。在工程实践中，真值往往是未知的，但数值仿真时可以给定真值；测量值通常由敏感器输出得到，往往含有测量误差；估计值主要是由估计方法确定得到的。

对于一个可测变量x，有两个方程，即

$$\widetilde{x}=x+\upsilon \tag{3.12}$$

$$\widetilde{x}=\hat{x}+e \tag{3.13}$$

式中：x，$\hat{x}$和$\widetilde{x}$分别为真值、估计值和测量值；υ为测量误差；e为残差。在实际应用中，往往遇到非线性参数模型，因此常采用非线性最小二乘法及其改进方法进行参数估计[4]。

假设有m个观测量，n个未知变量，且$m\geqslant n$，考虑非线性参数模型：

$$\widetilde{\boldsymbol{y}}=\boldsymbol{f}(\boldsymbol{x})+\boldsymbol{\upsilon} \tag{3.14}$$

式中：$\boldsymbol{f}(\cdot)$为连续可微的非线性函数；$\boldsymbol{x}\in\mathbb{R}^n$为待估变量；$\widetilde{\boldsymbol{y}}\in\mathbb{R}^m$为观测变量；$\boldsymbol{\upsilon}\in\mathbb{R}^m$为观测噪声。类似地，测量估计值和残差值可以写成

$$\hat{\boldsymbol{y}}=\boldsymbol{f}(\hat{\boldsymbol{x}}) \tag{3.15}$$

$$\boldsymbol{e}=\widetilde{\boldsymbol{y}}-\hat{\boldsymbol{y}}\equiv\Delta\boldsymbol{y} \tag{3.16}$$

因此，式（3.14）的测量模型可以用残差值表示成

$$\tilde{\boldsymbol{y}}=\boldsymbol{f}(\hat{\boldsymbol{x}})+\boldsymbol{e} \tag{3.17}$$

非线性最小二乘法是以误差的平方和最小为准则来估计静态模型参数的一种参数估计方法。因此，寻找最优估计 $\hat{\boldsymbol{x}}$ 使得损失函数 J 最小[5]：

$$J=\frac{1}{2}\boldsymbol{e}^{\mathrm{T}}\boldsymbol{W}\boldsymbol{e}=\frac{1}{2}[\tilde{\boldsymbol{y}}-\boldsymbol{f}(\hat{\boldsymbol{x}})]^{\mathrm{T}}\boldsymbol{W}[\tilde{\boldsymbol{y}}-\boldsymbol{f}(\hat{\boldsymbol{x}})] \tag{3.18}$$

式中：$\boldsymbol{W}\in\mathbb{R}^{m\times m}$ 为加权系数矩阵，用来赋予每个测量值的相对权重。

实际上，对于式（3.18），若想直接获得 $\hat{\boldsymbol{x}}$ 的解析闭合解很难，通常采用迭代法进行求解。假设 $\boldsymbol{x}_c$ 为已知的当前估计值，则修正后的估计值为

$$\hat{\boldsymbol{x}}=\boldsymbol{x}_c+\Delta\boldsymbol{x} \tag{3.19}$$

假设 $\Delta\boldsymbol{x}$ 是小量，把 $\boldsymbol{f}(\hat{\boldsymbol{x}})$ 在 $\boldsymbol{x}_c$ 处展成一阶泰勒表达式得

$$\boldsymbol{f}(\hat{\boldsymbol{x}})=\boldsymbol{f}(\boldsymbol{x}_c)+\boldsymbol{H}\Delta\boldsymbol{x} \tag{3.20}$$

其中，雅可比矩阵 $\boldsymbol{H}$ 为

$$\boldsymbol{H}=\left.\frac{\partial\boldsymbol{f}}{\partial\boldsymbol{x}}\right|_{\boldsymbol{x}_c} \tag{3.21}$$

于是，修正后的残差值可线性近似为

$$\Delta\boldsymbol{y}=\tilde{\boldsymbol{y}}-\boldsymbol{f}(\hat{\boldsymbol{x}})\approx\tilde{\boldsymbol{y}}-\boldsymbol{f}(\boldsymbol{x}_c)-\boldsymbol{H}\Delta\boldsymbol{x}=\Delta\boldsymbol{y}_c-\boldsymbol{H}\Delta\boldsymbol{x} \tag{3.22}$$

式中：$\Delta\boldsymbol{y}_c=\tilde{\boldsymbol{y}}-\boldsymbol{f}(\boldsymbol{x}_c)$ 为修正前的残差值。

于是，式（3.18）转换为求解 $\Delta\boldsymbol{x}$ 使得损失函数 J_p 最小：

$$J=\frac{1}{2}\Delta\boldsymbol{y}^{\mathrm{T}}\boldsymbol{W}\Delta\boldsymbol{y}\approx J_p\equiv\frac{1}{2}(\Delta\boldsymbol{y}_c-\boldsymbol{H}\Delta\boldsymbol{x})^{\mathrm{T}}\boldsymbol{W}(\Delta\boldsymbol{y}_c-\boldsymbol{H}\Delta\boldsymbol{x}) \tag{3.23}$$

利用线性最小二乘法求解可得

$$\Delta\boldsymbol{x}=(\boldsymbol{H}^{\mathrm{T}}\boldsymbol{W}\boldsymbol{H})^{-1}\boldsymbol{H}^{\mathrm{T}}\boldsymbol{W}\Delta\boldsymbol{y}_c \tag{3.24}$$

综上所述，非线性最小二乘算法流程如下：

步骤 1：选取初始值 $\boldsymbol{x}_c$，并给定容许误差 ε。

步骤 2：计算 $\Delta\boldsymbol{y}_c$，$\boldsymbol{H}$ 和 $\boldsymbol{W}$，利用式（3.24）计算 $\Delta\boldsymbol{x}$。

步骤 3：检查是否满足终止条件。若 $\|\Delta\boldsymbol{x}\|<\varepsilon$，迭代终止；否则，转步骤 4。

步骤 4：利用 $\Delta\boldsymbol{x}$ 修正当前估计值 $\boldsymbol{x}_c$，即

$$\boldsymbol{x}_c=\boldsymbol{x}_c+\Delta\boldsymbol{x} \tag{3.25}$$

步骤 5：返回步骤 2。该过程是一个迭代求解过程，直到算法收敛或者是达到预定迭代次数才终止。

图 3.1 给出了非线性最小二乘算法流程。

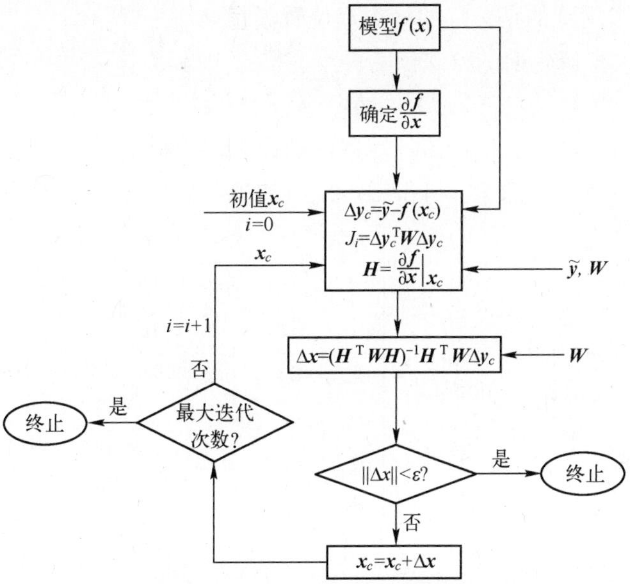

图 3.1　非线性最小二乘法流程

3.2.2　相对位姿确定方法

针对含噪声的 VISNAV 观测模型，本节采用非线性最小二乘法求解航天器相对位置和姿态。对应于式（3.9），标志点的像平面坐标估计方程为

$$\hat{\boldsymbol{y}}=\hat{\boldsymbol{c}}=\boldsymbol{f}(\hat{\boldsymbol{x}}) \tag{3.26}$$

式中：

$$\hat{\boldsymbol{c}}=\begin{bmatrix}\hat{\boldsymbol{c}}_1\\ \hat{\boldsymbol{c}}_2\\ \vdots\\ \hat{\boldsymbol{c}}_N\end{bmatrix}\equiv\begin{bmatrix}\hat{\alpha}_1\\ \hat{\gamma}_1\\ \hat{\alpha}_2\\ \hat{\gamma}_2\\ \vdots\\ \hat{\alpha}_N\\ \hat{\gamma}_N\end{bmatrix}_{2N\times1},\quad \hat{\boldsymbol{x}}=\begin{bmatrix}\hat{\boldsymbol{s}}^c\\ \hat{\boldsymbol{p}}_{d/c}\end{bmatrix}\equiv\begin{bmatrix}\hat{x}^c\\ \hat{y}^c\\ \hat{z}^c\\ \hat{p}_1\\ \hat{p}_2\\ \hat{p}_3\end{bmatrix}_{6\times1} \tag{3.27}$$

式（3.26）中第 i 个标志点的像平面坐标估计值可写为

$$\hat{\boldsymbol{c}}_i=\begin{bmatrix}\hat{\alpha}_i\\ \hat{\gamma}_i\end{bmatrix}=\boldsymbol{f}(\hat{\boldsymbol{x}},\boldsymbol{\mathcal{X}}_i),\quad i=1,2,\cdots,N \tag{3.28}$$

式（3.28）中用到的是标称的标志点位置矢量 $\boldsymbol{\chi}_i$。

因此，可得修正前的残差值为

$$\Delta \boldsymbol{y}_c=\tilde{\boldsymbol{c}}-\hat{\boldsymbol{c}}_c \tag{3.29}$$

式中：$\hat{\boldsymbol{c}}_c$ 为由当前状态估计值计算得到的像平面坐标估值。

经过一系列数学推导，可得雅可比矩阵为

$$\boldsymbol{H}=\frac{\partial \boldsymbol{f}}{\partial \boldsymbol{x}}=\begin{bmatrix}\dfrac{\partial \boldsymbol{c}}{\partial \boldsymbol{s}^c} & \dfrac{\partial \boldsymbol{c}}{\partial \boldsymbol{p}_{d/c}}\end{bmatrix}=\begin{bmatrix}\dfrac{\partial \boldsymbol{c}_1}{\partial x^c} & \dfrac{\partial \boldsymbol{c}_1}{\partial y^c} & \dfrac{\partial \boldsymbol{c}_1}{\partial z^c} & \dfrac{\partial \boldsymbol{c}_1}{\partial p_1} & \dfrac{\partial \boldsymbol{c}_1}{\partial p_2} & \dfrac{\partial \boldsymbol{c}_1}{\partial p_3} \\ \dfrac{\partial \boldsymbol{c}_2}{\partial x^c} & \dfrac{\partial \boldsymbol{c}_2}{\partial y^c} & \dfrac{\partial \boldsymbol{c}_2}{\partial z^c} & \dfrac{\partial \boldsymbol{c}_2}{\partial p_1} & \dfrac{\partial \boldsymbol{c}_2}{\partial p_2} & \dfrac{\partial \boldsymbol{c}_2}{\partial p_3} \\ \vdots & \vdots & \vdots & \vdots & \vdots & \vdots \\ \dfrac{\partial \boldsymbol{c}_N}{\partial x^c} & \dfrac{\partial \boldsymbol{c}_N}{\partial y^c} & \dfrac{\partial \boldsymbol{c}_N}{\partial z^c} & \dfrac{\partial \boldsymbol{c}_N}{\partial p_1} & \dfrac{\partial \boldsymbol{c}_N}{\partial p_2} & \dfrac{\partial \boldsymbol{c}_N}{\partial p_3}\end{bmatrix} \tag{3.30}$$

其中，

$$\frac{\partial \boldsymbol{c}_i}{\partial x^c}=\begin{bmatrix}\dfrac{\partial \alpha_i}{\partial x^c} \\ \dfrac{\partial \gamma_i}{\partial x^c}\end{bmatrix}=\begin{bmatrix}\dfrac{f[A_{11}A_{32}(Y_i-y^c)+A_{11}A_{33}(Z_i-z^c)-A_{12}A_{31}(Y_i-y^c)-A_{13}A_{31}(Z_i-z^c)]}{[A_{31}(X_i-x^c)+A_{32}(Y_i-y^c)+A_{33}(Z_i-z^c)]^2} \\ \dfrac{f[A_{21}A_{32}(Y_i-y^c)+A_{21}A_{33}(Z_i-z^c)-A_{22}A_{31}(Y_i-y^c)-A_{23}A_{31}(Z_i-z^c)]}{[A_{31}(X_i-x^c)+A_{32}(Y_i-y^c)+A_{33}(Z_i-z^c)]^2}\end{bmatrix} \tag{3.31}$$

$$\frac{\partial \boldsymbol{c}_i}{\partial y^c}=\begin{bmatrix}\dfrac{\partial \alpha_i}{\partial y^c} \\ \dfrac{\partial \gamma_i}{\partial y^c}\end{bmatrix}=\begin{bmatrix}\dfrac{f[A_{12}A_{31}(X_i-x^c)+A_{12}A_{33}(Z_i-z^c)-A_{11}A_{32}(X_i-x^c)-A_{13}A_{32}(Z_i-z^c)]}{[A_{31}(X_i-x^c)+A_{32}(Y_i-y^c)+A_{33}(Z_i-z^c)]^2} \\ \dfrac{f[A_{22}A_{31}(X_i-x^c)+A_{22}A_{33}(Z_i-z^c)-A_{21}A_{32}(X_i-x^c)-A_{23}A_{32}(Z_i-z^c)]}{[A_{31}(X_i-x^c)+A_{32}(Y_i-y^c)+A_{33}(Z_i-z^c)]^2}\end{bmatrix} \tag{3.32}$$

$$\frac{\partial \boldsymbol{c}_i}{\partial z^c}=\begin{bmatrix}\dfrac{\partial \alpha_i}{\partial z^c} \\ \dfrac{\partial \gamma_i}{\partial z^c}\end{bmatrix}=\begin{bmatrix}\dfrac{f[A_{13}A_{31}(X_i-x^c)+A_{13}A_{32}(Y_i-y^c)-A_{11}A_{33}(X_i-x^c)-A_{12}A_{33}(Y_i-y^c)]}{[A_{31}(X_i-x^c)+A_{32}(Y_i-y^c)+A_{33}(Z_i-z^c)]^2} \\ \dfrac{f[A_{23}A_{31}(X_i-x^c)+A_{23}A_{32}(Y_i-y^c)-A_{21}A_{33}(X_i-x^c)-A_{22}A_{33}(Y_i-y^c)]}{[A_{31}(X_i-x^c)+A_{32}(Y_i-y^c)+A_{33}(Z_i-z^c)]^2}\end{bmatrix} \tag{3.33}$$

进一步，可将式（3.31）~式（3.33）写成

$$\frac{\partial \boldsymbol{c}_i}{\partial \boldsymbol{s}^c}=\frac{f}{[A_{31}(X_i-x^c)+A_{32}(Y_i-y^c)+A_{33}(Z_i-z^c)]^2}\cdot \boldsymbol{M}\cdot \boldsymbol{\Lambda} \tag{3.34}$$

其中，

$$\boldsymbol{M}=\begin{bmatrix}0 & \boldsymbol{M}_{12} & \boldsymbol{M}_{13} & \boldsymbol{M}_{14} & 0 & \boldsymbol{M}_{16} & \boldsymbol{M}_{17} & \boldsymbol{M}_{18} & 0\\ 0 & \boldsymbol{M}_{22} & \boldsymbol{M}_{23} & \boldsymbol{M}_{24} & 0 & \boldsymbol{M}_{26} & \boldsymbol{M}_{27} & \boldsymbol{M}_{28} & 0\end{bmatrix}_{2\times 9} \tag{3.35}$$

$$\boldsymbol{M}_{12}=(A_{11}A_{32}-A_{12}A_{31}),\quad \boldsymbol{M}_{13}=(A_{11}A_{33}-A_{13}A_{31})$$
$$\boldsymbol{M}_{14}=(A_{12}A_{31}-A_{11}A_{32}),\quad \boldsymbol{M}_{16}=(A_{12}A_{33}-A_{13}A_{32})$$
$$\boldsymbol{M}_{17}=(A_{13}A_{31}-A_{11}A_{33}),\quad \boldsymbol{M}_{18}=(A_{13}A_{32}-A_{12}A_{33})$$
$$\boldsymbol{M}_{22}=(A_{21}A_{32}-A_{22}A_{31}),\quad \boldsymbol{M}_{23}=(A_{21}A_{33}-A_{23}A_{31})$$
$$\boldsymbol{M}_{24}=(A_{22}A_{31}-A_{21}A_{32}),\quad \boldsymbol{M}_{26}=(A_{22}A_{33}-A_{23}A_{32})$$
$$\boldsymbol{M}_{27}=(A_{23}A_{31}-A_{21}A_{33}),\quad \boldsymbol{M}_{28}=(A_{23}A_{32}-A_{22}A_{33})$$

$$\boldsymbol{\Lambda}=\begin{bmatrix}(X_i-x^c) & 0 & 0\\ (Y_i-y^c) & 0 & 0\\ (Z_i-z^c) & 0 & 0\\ 0 & (X_i-x^c) & 0\\ 0 & (Y_i-y^c) & 0\\ 0 & (Z_i-z^c) & 0\\ 0 & 0 & (X_i-x^c)\\ 0 & 0 & (Y_i-y^c)\\ 0 & 0 & (Z_i-z^c)\end{bmatrix}_{9\times 3} \tag{3.36}$$

另外，对于偏导数矩阵$\partial \boldsymbol{c}/\partial \boldsymbol{p}_{d/c}$，根据偏微分的链式法则，推导可得

$$\frac{\partial \boldsymbol{c}_i}{\partial \boldsymbol{p}_{d/c}}=\frac{\partial \boldsymbol{c}_i}{\partial \boldsymbol{b}_i}\frac{\partial \boldsymbol{b}_i}{\partial \boldsymbol{p}_{d/c}},\quad i=1,2,\cdots,N \tag{3.37}$$

式中：$\boldsymbol{b}_i=\boldsymbol{A}_c^d(\boldsymbol{p}_{d/c})\boldsymbol{r}_i$ 为第 i 个视线单位观测矢量，且由式（2.35）计算可得

$$\frac{\partial \boldsymbol{c}_i}{\partial \boldsymbol{b}_i}=\begin{bmatrix}\dfrac{-f}{b_{iz}} & 0 & f\dfrac{b_{ix}}{b_{iz}^2}\\ 0 & \dfrac{-f}{b_{iz}} & f\dfrac{b_{iy}}{b_{iz}^2}\end{bmatrix} \tag{3.38}$$

对于偏导数矩阵$\partial \boldsymbol{b}_i/\partial \boldsymbol{p}_{d/c}$，计算可得

$$\begin{aligned}\frac{\partial \boldsymbol{b}_i}{\partial \boldsymbol{p}_{d/c}}&=\frac{\partial \boldsymbol{A}_c^d(\boldsymbol{p}_{d/c})\boldsymbol{r}_i}{\partial \boldsymbol{p}_{d/c}}\\ &=\frac{4}{(1+\boldsymbol{p}_{d/c}^{\mathrm{T}}\boldsymbol{p}_{d/c})^2}[\boldsymbol{A}_c^d(\boldsymbol{p}_{d/c})\boldsymbol{r}_i\times]\{(1-\boldsymbol{p}_{d/c}^{\mathrm{T}}\boldsymbol{p}_{d/c})\boldsymbol{I}_{3\times 3}-2[\boldsymbol{p}_{d/c}\times]+2\boldsymbol{p}_{d/c}\boldsymbol{p}_{d/c}^{\mathrm{T}}\}\end{aligned} \tag{3.39}$$

至此，已计算得到雅可比矩阵 $\boldsymbol{H}$ 中的所有元素，可以采用 3.2.1 节中的非线性最小二乘法求解相对位置和姿态。

3.2.3 加权矩阵 W 的选取

本节将详细推导观测误差噪声阵 $\boldsymbol{R}$，并用于前述的非线性最小二乘法中。这里，观测误差噪声阵 $\boldsymbol{R}$ 综合考虑了标志点位置误差噪声和 VISNAV 系统测量噪声两方面影响。根据前文的推导，可知含噪声的 VISNAV 观测模型为[6]

$$\tilde{\boldsymbol{c}}=\boldsymbol{f}(\boldsymbol{x},\boldsymbol{v}_1,\boldsymbol{v}_2)=\boldsymbol{f}(\boldsymbol{x})+\boldsymbol{w}_1(\boldsymbol{x},\boldsymbol{v}_1)+\boldsymbol{w}_2(\boldsymbol{v}_2) \tag{3.40}$$

其中，$\boldsymbol{w}_1(\boldsymbol{x},\boldsymbol{v}_1)$ 表示由标志点位置误差噪声 $\boldsymbol{v}_1$ 造成的像平面坐标测量噪声，该误差项是以非线性方式通过透视投影方程组引入的。因此，对该非线性噪声项一阶泰勒展开可得

$$\begin{aligned}\tilde{\boldsymbol{c}}_i&=\boldsymbol{f}(\boldsymbol{x},\boldsymbol{\mathcal{X}}_i,\boldsymbol{v}_{1i})+\boldsymbol{v}_{2i}\\&=\boldsymbol{f}(\boldsymbol{x},\boldsymbol{\mathcal{X}}_i)+\boldsymbol{M}_i\boldsymbol{v}_{1i}+\boldsymbol{v}_{2i}\end{aligned} \tag{3.41}$$

式中：

$$\boldsymbol{M}_i=\frac{\partial\boldsymbol{c}_i}{\partial\boldsymbol{\mathcal{X}}_i}=\begin{bmatrix}\dfrac{\partial\boldsymbol{\alpha}_i}{\partial X_i} & \dfrac{\partial\boldsymbol{\alpha}_i}{\partial Y_i} & \dfrac{\partial\boldsymbol{\alpha}_i}{\partial Z_i}\\[2ex] \dfrac{\partial\boldsymbol{\gamma}_i}{\partial X_i} & \dfrac{\partial\boldsymbol{\gamma}_i}{\partial Y_i} & \dfrac{\partial\boldsymbol{\gamma}_i}{\partial Z_i}\end{bmatrix}_{2\times 3} \tag{3.42}$$

考虑到式（3.6）和式（3.7）中矢量 $\boldsymbol{\mathcal{X}}_i\equiv[X_i,Y_i,Z_i]^{\mathrm{T}}$ 与 $\boldsymbol{s}^c\equiv[x^c,y^c,z^c]^{\mathrm{T}}$ 之间的对称性，可得

$$\frac{\partial\boldsymbol{c}_i}{\partial\boldsymbol{\mathcal{X}}_i}=-\frac{\partial\boldsymbol{c}_i}{\partial\boldsymbol{s}^c}=\frac{-f}{[A_{31}(X_i-x^c)+A_{32}(Y_i-y^c)+A_{33}(Z_i-z^c)]^2}\cdot\boldsymbol{M}\cdot\boldsymbol{\Lambda} \tag{3.43}$$

这里，矩阵 $\boldsymbol{M}$ 和 $\boldsymbol{\Lambda}$ 已在前文中进行了定义。

因此，观测误差噪声阵 $\boldsymbol{R}$ 的计算公式如下：

$$\boldsymbol{R}=\boldsymbol{\mathcal{R}}_1+\boldsymbol{\mathcal{R}}_2 \tag{3.44}$$

其中，

$$\boldsymbol{\mathcal{R}}_1=\begin{bmatrix}R_1 & 0 & \cdots & 0 & 0\\ 0 & R_2 & 0 & \cdots & 0\\ \vdots & 0 & & & \vdots\\ 0 & \vdots & & & 0\\ 0 & 0 & \cdots & 0 & R_N\end{bmatrix}_{2N\times 2N} \tag{3.45}$$

$$R_i=\sigma_1^2\boldsymbol{M}_i\boldsymbol{M}_i^{\mathrm{T}},\quad i=1,2,\cdots,N \tag{3.46}$$

$$\boldsymbol{\mathcal{R}}_2=\sigma_2^2\boldsymbol{I}_{2N\times 2N} \tag{3.47}$$

需要注意的是，在上述推导过程中，假设噪声 $\boldsymbol{v}_1$ 和 $\boldsymbol{v}_2$ 是不相关的，即 $\boldsymbol{E}\{\boldsymbol{v}_1\boldsymbol{v}_2^{\mathrm{T}}\}=\boldsymbol{E}\{\boldsymbol{v}_2\boldsymbol{v}_1^{\mathrm{T}}\}=\boldsymbol{0}$。

由高斯-可尔可夫定理[7]可知，最优加权系数矩阵为

$$\boldsymbol{W}=\boldsymbol{R}^{-1} \tag{3.48}$$

可以证明，此时估计误差的方差最小，即达到 Cramér-Rao 下界（Cramér-Rao Lower Bound，CRLB）。该值是根据信息论准则确定的，与估计方法没有关系，可作为衡量估计方法性能的标准。通常，将蒙特卡洛仿真所得到的参数的估计方差同 CRLB 进行比较，越接近 CRLB 说明估计方法越有效[8]。

这里，定义 $\boldsymbol{P}_s$ 为蒙特卡洛仿真所得到的参数的估计误差方差阵，$\boldsymbol{P}_A$ 为理论估计误差方差阵，即[9]

$$\boldsymbol{P}_A=\boldsymbol{E}\{(\hat{\boldsymbol{x}}-\boldsymbol{x})(\hat{\boldsymbol{x}}-\boldsymbol{x})^{\mathrm{T}}\}=(\boldsymbol{H}^{\mathrm{T}}\boldsymbol{R}^{-1}\boldsymbol{H})^{-1} \tag{3.49}$$

此时，计算式（3.49）中的 $\boldsymbol{H}$ 阵用到的是相对位置和姿态的真值信息。理论上，这两个矩阵大小相当，而且随着蒙特卡洛仿真次数的增加，$\boldsymbol{P}_s$ 更加趋近于 $\boldsymbol{P}_A$。

3.3　仿真实验与分析

本节对所设计的相对位姿确定方法进行数值仿真验证，分析标志点位置误差噪声对相对位置和姿态测量精度的影响。数值仿真分为静态仿真和动态仿真两部分。其中，静态仿真是采用蒙特卡洛法对某一给定的相对位置和姿态进行解算，动态仿真是模拟一个交会场景，实时求解主从星之间的相对位置和姿态。

3.3.1　静态仿真

本节针对基于非线性最小二乘法的相对位姿确定方法进行蒙特卡洛仿真分析。表 3.1 列出了主星上 6 个标志点的标称位置坐标。简便起见，假设焦距 $f=1\mathrm{m}$，详细仿真参数如表 3.2 所示，蒙特卡洛仿真 1000 次，仿真流程如图 3.2 所示。

状态初值选取为 $\hat{\boldsymbol{x}}=[1,1,1,0,0,0]^{\mathrm{T}}$，采用非线性最小二乘法求解航天器相对位置和姿态，蒙特卡洛仿真结果如图 3.3～图 3.6 所示。图 3.3 和图 3.4 分别给出了相对位置和姿态估计误差与 3σ 界。这里，3σ 界是由式（3.49）计算得到的。可以看出，估计误差基本上全部控制在 3σ 界内，与理论相符。

系统的相对定位精度为 0.07m，相对定姿精度为 0.06°。图 3.5 和图 3.6 分别给出了相对位置和姿态的估计误差散布。

表 3.1　标志点标称位置

序号	X/m	Y/m	Z/m
1	0.5	0.5	0.0
2	-0.5	-0.5	0.0
3	-0.5	0.5	0.0
4	0.5	-0.5	0.0
5	0.2	0.5	0.1
6	0.0	0.2	-0.1

表 3.2　仿真参数

仿 真 参 数	数　值
相对位置真值	$\boldsymbol{s}^c=[30,30,50]^{\mathrm{T}}\mathrm{m}$
相对姿态真值	$\boldsymbol{p}_{d/c}=[0.1,0.1,0.1]^{\mathrm{T}}$
噪声标准差	$\sigma_1=2\times10^{-4}\mathrm{m}$，$\sigma_2=1\times10^{-5}\mathrm{m}$
焦距	$f=1\mathrm{m}$

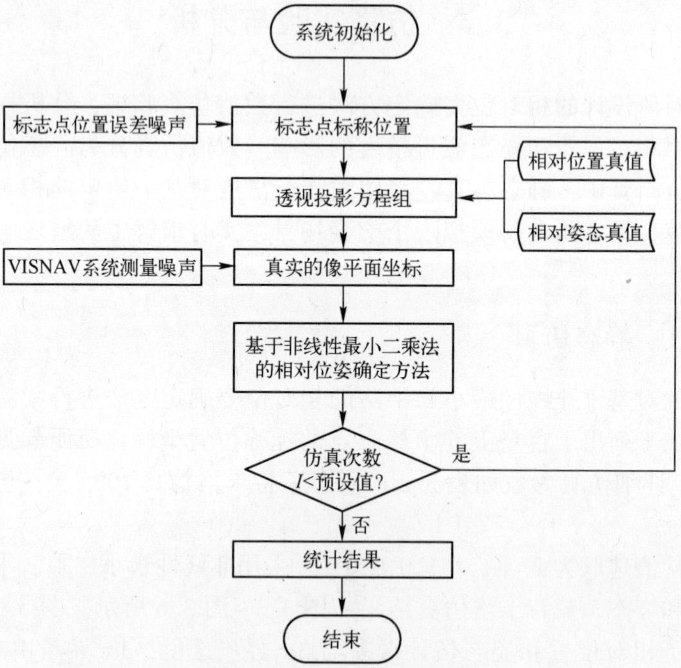

图 3.2　蒙特卡洛仿真流程

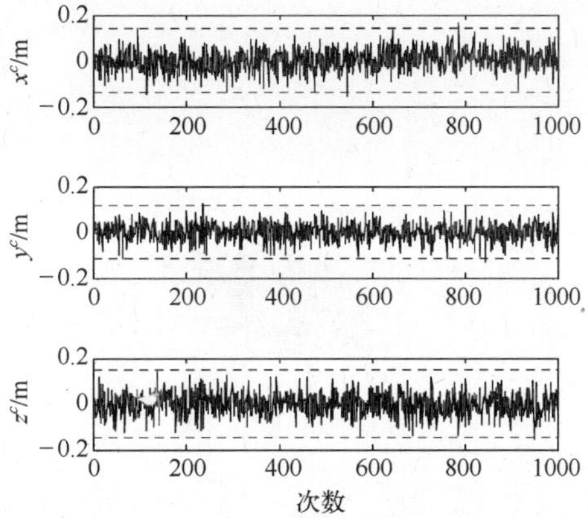

图 3.3　相对位置估计误差与 3σ 界（$\sigma_1 = 0.2\text{mm}$）

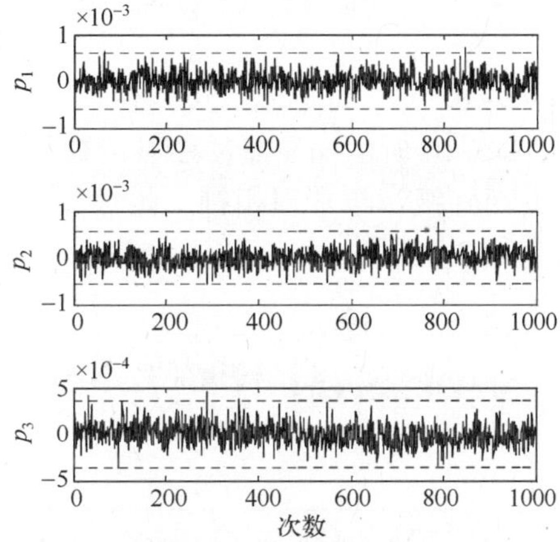

图 3.4　相对姿态估计误差与 3σ 界（$\sigma_1 = 0.2\text{mm}$）

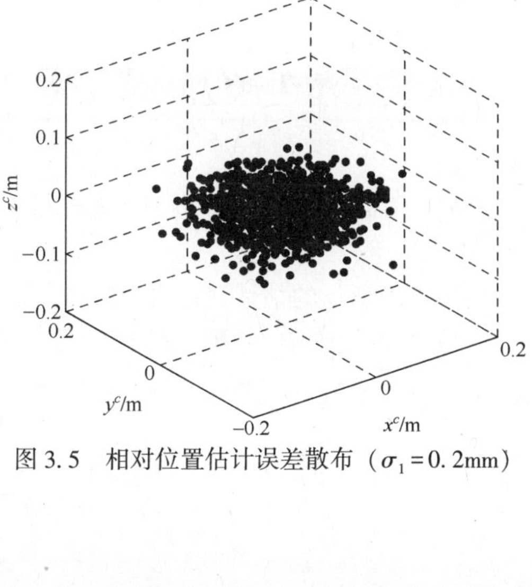

图 3.5　相对位置估计误差散布（$\sigma_1 = 0.2\text{mm}$）

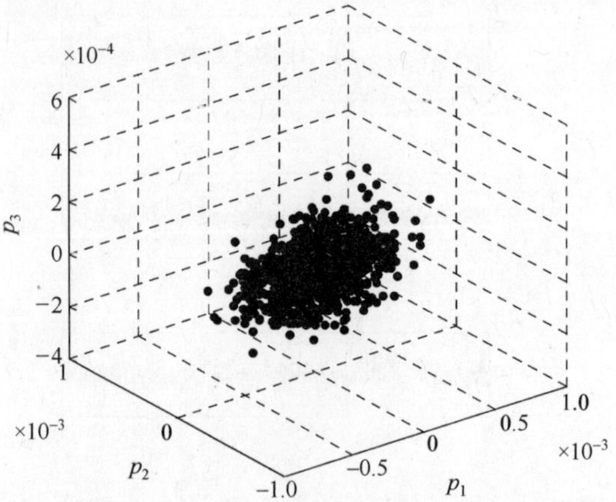

图 3.6　相对姿态估计误差散布（$\sigma_1 = 0.2$mm）

进一步，选取 $\sigma_1 = 2\times10^{-3}$m，将标志点位置误差噪声放大 10 倍，蒙特卡洛仿真结果如图 3.7～图 3.10 所示。显然，相对位置和姿态估计精度随着标志点位置误差噪声放大而降级，与常理相符。此时，系统的相对定位精度为 0.3m，相对定姿精度为 0.3°。

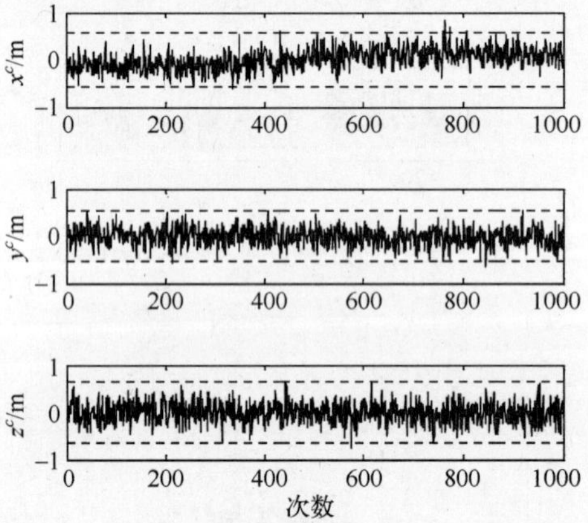

图 3.7　相对位置估计误差与 3σ 界（$\sigma_1 = 2$mm）

表 3.3 列出了两种仿真情形下的估计误差方差阵 $\boldsymbol{P}_A$ 和 $\boldsymbol{P}_s$。分析可知，蒙特卡洛仿真得到的估计误差方差阵 $\boldsymbol{P}_s$ 与理论估计误差方差阵 $\boldsymbol{P}_A$ 大小相当，与理论相符。

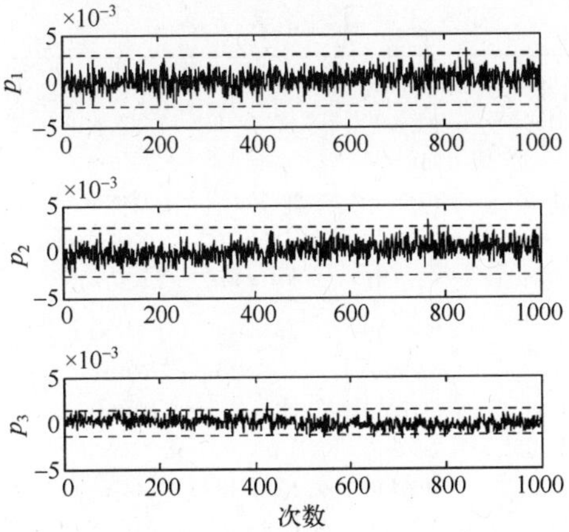

图 3.8　相对姿态估计误差与 3σ 界（$\sigma_1 = 2\text{mm}$）

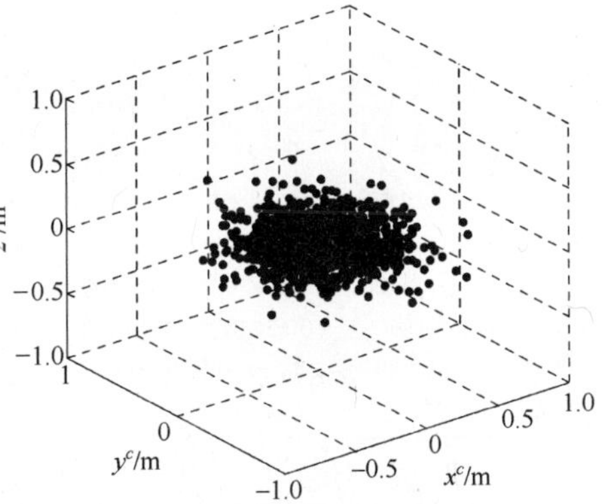

图 3.9　相对位置估计误差散布图（$\sigma_1 = 2\text{mm}$）

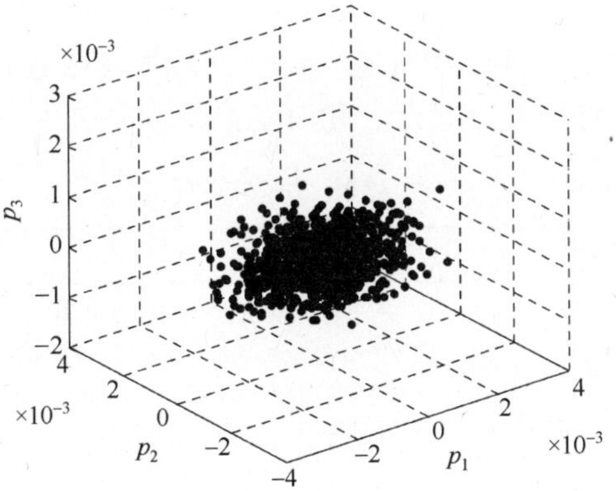

图 3.10　相对姿态估计误差散布图（$\sigma_1 = 2\text{mm}$）

表 3.3　估计误差方差阵比较

情形一 $\sigma_1=2\times10^{-4}$m	$\boldsymbol{P}_A=\begin{bmatrix} 2.1364\times10^{-3} & -8.0051\times10^{-4} & -6.2774\times10^{-4} & 4.8228\times10^{-6} & 7.2732\times10^{-6} & -4.1279\times10^{-6} \\ -8.0051\times10^{-4} & 1.4447\times10^{-3} & -7.8017\times10^{-4} & -7.4082\times10^{-6} & -3.6177\times10^{-7} & 1.0037\times10^{-6} \\ -6.2774\times10^{-4} & -7.8017\times10^{-4} & 2.3484\times10^{-3} & 5.3987\times10^{-6} & -6.6124\times10^{-6} & 2.8641\times10^{-6} \\ 4.8228\times10^{-6} & -7.4082\times10^{-6} & 5.3987\times10^{-6} & 4.2242\times10^{-8} & 2.9869\times10^{-9} & -2.6291\times10^{-9} \\ 7.2732\times10^{-6} & -3.6177\times10^{-7} & -6.6124\times10^{-6} & 2.9869\times10^{-9} & 3.6203\times10^{-8} & -1.4377\times10^{-8} \\ -4.1279\times10^{-6} & 1.0037\times10^{-6} & 2.8641\times10^{-6} & -2.6291\times10^{-9} & -1.4377\times10^{-8} & 1.3639\times10^{-8} \end{bmatrix}$
	$\boldsymbol{P}_s=\begin{bmatrix} 2.1591\times10^{-3} & -8.1690\times10^{-4} & -6.1047\times10^{-4} & 4.9211\times10^{-6} & 7.2788\times10^{-6} & -4.1833\times10^{-6} \\ -8.1690\times10^{-4} & 1.5027\times10^{-3} & -8.1747\times10^{-4} & -7.6950\times10^{-6} & -3.2100\times10^{-7} & 1.0110\times10^{-6} \\ -6.1047\times10^{-4} & -8.1747\times10^{-4} & 2.3525\times10^{-3} & 5.6048\times10^{-6} & -6.5270\times10^{-6} & 2.9078\times10^{-6} \\ 4.9211\times10^{-6} & -7.6950\times10^{-6} & 5.6048\times10^{-6} & 4.3756\times10^{-8} & 2.8431\times10^{-9} & -2.6002\times10^{-9} \\ 7.2788\times10^{-6} & -3.2100\times10^{-7} & -6.5270\times10^{-6} & 2.8431\times10^{-9} & 3.5898\times10^{-8} & -1.4535\times10^{-8} \\ -4.1833\times10^{-6} & 1.0110\times10^{-6} & 2.9078\times10^{-6} & -2.6002\times10^{-9} & -1.4535\times10^{-8} & 1.3888\times10^{-8} \end{bmatrix}$
情形二 $\sigma_1=2\times10^{-3}$m	$\boldsymbol{P}_A=\begin{bmatrix} 3.7176\times10^{-2} & -1.6114\times10^{-2} & -1.3232\times10^{-2} & 9.0741\times10^{-5} & 1.3307\times10^{-4} & -7.0370\times10^{-5} \\ -1.6114\times10^{-2} & 3.4237\times10^{-2} & -1.5473\times10^{-2} & -1.6396\times10^{-4} & -1.3257\times10^{-6} & 2.8483\times10^{-5} \\ -1.3232\times10^{-2} & -1.5473\times10^{-2} & 4.8891\times10^{-2} & 1.0094\times10^{-4} & -1.4878\times10^{-4} & 4.4401\times10^{-5} \\ 9.0741\times10^{-5} & -1.6396\times10^{-4} & 1.0094\times10^{-4} & 8.6517\times10^{-7} & 2.5648\times10^{-8} & -9.0303\times10^{-8} \\ 1.3307\times10^{-4} & -1.3257\times10^{-6} & -1.4878\times10^{-4} & 2.5648\times10^{-8} & 7.5640\times10^{-7} & -2.2984\times10^{-7} \\ -7.0370\times10^{-5} & 2.8483\times10^{-5} & 4.4401\times10^{-5} & -9.0303\times10^{-8} & -2.2984\times10^{-7} & 2.3587\times10^{-7} \end{bmatrix}$
	$\boldsymbol{P}_s=\begin{bmatrix} 4.6192\times10^{-2} & -1.7598\times10^{-2} & -1.4970\times10^{-2} & 9.8194\times10^{-5} & 1.5314\times10^{-4} & -9.9978\times10^{-5} \\ -1.7598\times10^{-2} & 3.5705\times10^{-2} & -1.5815\times10^{-2} & -1.6911\times10^{-4} & -1.1854\times10^{-6} & 3.5772\times10^{-5} \\ -1.4970\times10^{-2} & -1.5815\times10^{-2} & 4.8022\times10^{-2} & 1.0165\times10^{-4} & -1.4846\times10^{-4} & 5.1945\times10^{-5} \\ 9.8194\times10^{-5} & -1.6911\times10^{-4} & 1.0165\times10^{-4} & 8.8419\times10^{-7} & 3.1005\times10^{-8} & -1.2221\times10^{-7} \\ 1.5314\times10^{-4} & -1.1854\times10^{-6} & -1.4846\times10^{-4} & 3.1005\times10^{-8} & 7.8803\times10^{-7} & -3.0020\times10^{-7} \\ -9.9978\times10^{-5} & 3.5772\times10^{-5} & 5.1945\times10^{-5} & -1.2221\times10^{-7} & -3.0020\times10^{-7} & 3.3298\times10^{-7} \end{bmatrix}$

3.3.2　动态仿真

本节模拟了一个交会场景，采用所设计的相对位姿确定方法实时求解主从星之间的相对位置和姿态。详细仿真参数如表 3.4 所示，主从星之间的相对运动轨迹（包括平动和转动轨迹）可依据表中参数进行数值模拟。其中，相对平动运动轨迹如图 3.11 所示，相对姿态运动轨迹由式（3.4）传播得到。

表 3.4　仿真参数

仿真参数	数　值
相对位置真值	$x^c=30-(30/1050)t(\mathrm{m})$ $y^c=30-(30/1050)t(\mathrm{m})$ $z^c=50\exp[(-1/300)t](\mathrm{m})$
初始相对姿态真值	$\boldsymbol{p}_{d/c}(0)=[0.02,-0.02,0.02]^{\mathrm{T}}$
相对角速度	$\boldsymbol{\omega}_{d/c}=[0,0.0011,0]^{\mathrm{T}}(\mathrm{rad/s})$

续表

仿真参数	数　值
噪声标准差	$\sigma_1=2\times10^{-3}$m，$\sigma_2=1\times10^{-5}$m
焦距	$f=1$m
仿真时间	$t=1000$s

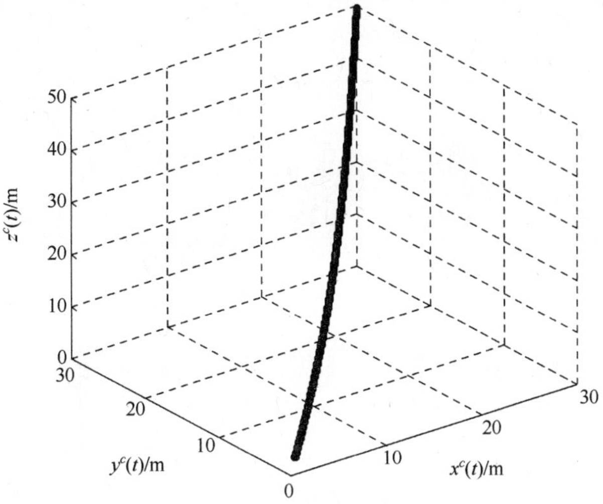

图 3.11　模拟相对平动运动轨迹

状态初值选取为 $\hat{\boldsymbol{x}}(0)=[1,1,1,0,0,0]^{\mathrm{T}}$，采用非线性最小二乘法实时求解主从星之间的相对位置和姿态，前一时刻解算出的状态值作为后一时刻的迭代初值。仿真结果如图 3.12~图 3.17 所示。图 3.12 和图 3.13 分别给

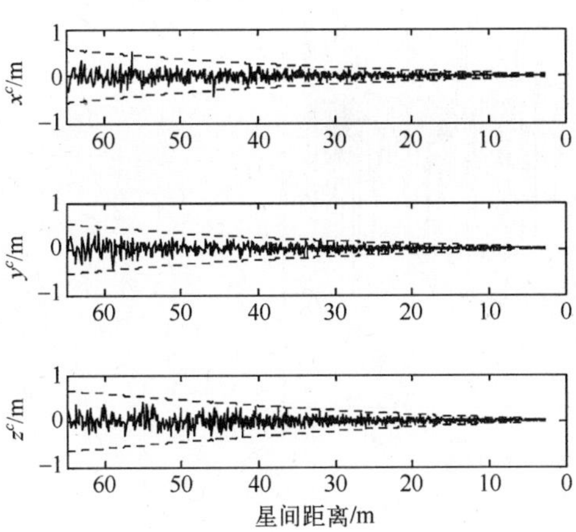

图 3.12　相对位置估计误差与 3σ 界

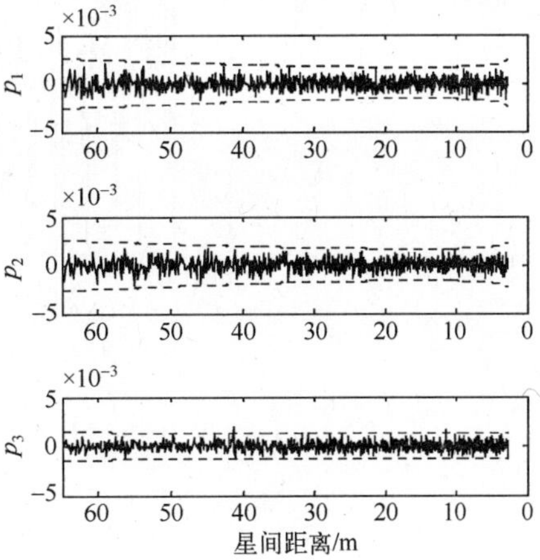

图 3.13　相对姿态估计误差与 3σ 界

出了相对位置和姿态估计误差与3σ界。图3.14和图3.15分别给出了相对位置和姿态估计误差范数。可以看出，一开始从星距离主星约65m，随着从星不断逼近主星，相对位置和姿态估计精度越来越高，相对位置估计误差由最初的0.45m下降到0.01m，相对姿态估计误差由最初的0.4°下降到0.1°。图3.16给出了估计误差方差阵的条件数，可以看出，随着主从星之间距离的接近，系统的可观测性越来越好。图3.17给出了非线性最小二乘算法的迭代次数。

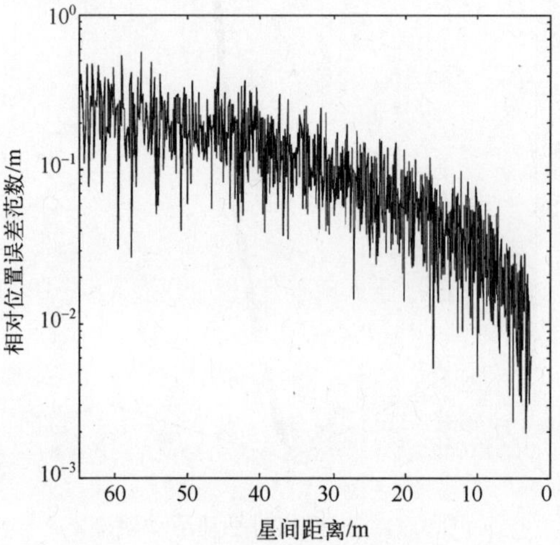

图3.14 相对位置估计误差范数

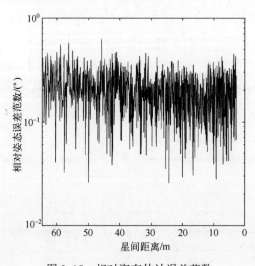

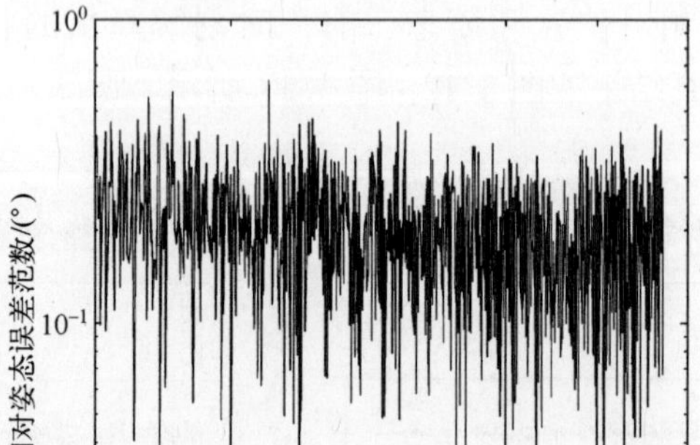

图3.15 相对姿态估计误差范数

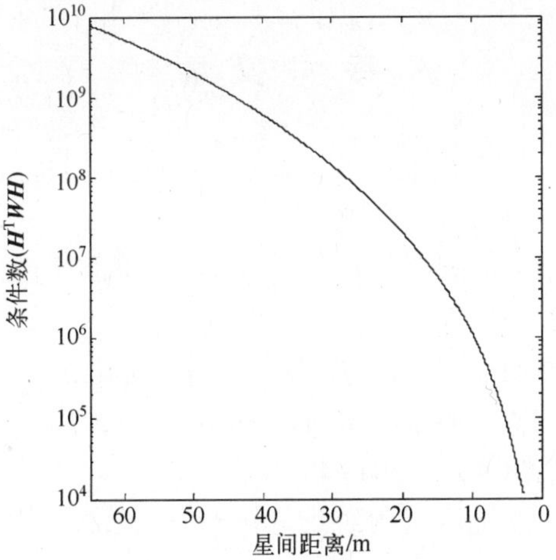

图 3.16　估计误差方差阵的条件数

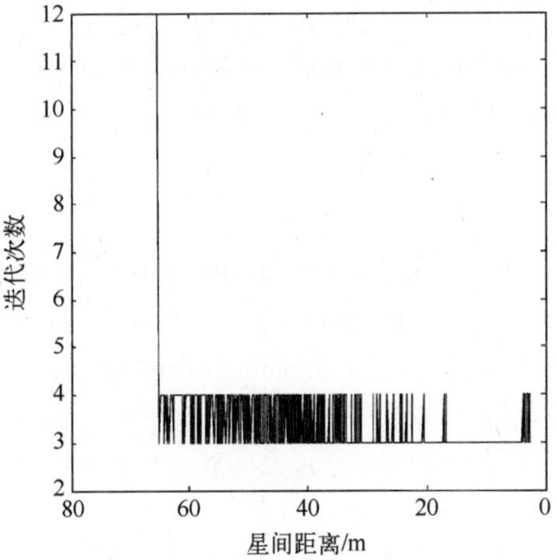

图 3.17　非线性最小二乘法的迭代次数

仿真结果表明，标志点位置误差噪声对相对位置和姿态测量精度影响较大；随着主从星之间距离的接近，系统的可观测性越来越好，相对位置和姿态估计精度也越来越高。

需要指出的是，这里求解得到的相对位置矢量是在主星体系中进行描述的，利用姿态阵 $\boldsymbol{A}_H^c$ 可将相对位置矢量 $\boldsymbol{s}^c$ 转换到 LVLH 系，结合所求的相对姿态阵 $\boldsymbol{A}_c^d$ 可进一步求解得到姿态阵 $\boldsymbol{A}_H^d$，为下一章设计的导航滤波器提供初值。

参考文献

[1] WERTZ J R. Spacecraft attitude determination and control [M]. Dordrecht: Kluwer Academic Publishers, 1978.

[2] SHUSTER M D. A survey of attitude representations [J]. Journal of the Astronautical Sciences, 1993, 41 (4): 439-517.

[3] CRASSIDIS J L, ALONSO R, JUNKINS J L. Optimal attitude and position determination from line-of-sight measurements [J]. Journal of the Astronautical Sciences, 2000, 48 (2): 391-408.

[4] 张金槐, 蔡洪. 飞行器试验统计学 [M]. 长沙: 国防科技大学出版社, 1995.

[5] MAYBECK P S. Stochastic models, estimation, and control: Vol. 2 [M]. Arlington: Navtech Book and Software Store, 1994.

[6] CRASSIDIS J, JUNKINS J. Optimal estimation of dynamic systems [M]. Boca Raton: Chapman & Hall/CRC Press, 2004.

[7] CRAMER H. Mathematical methods of statistics [M]. Princeton: Princeton University Press, 1946.

[8] SAGE A P, MELSA J L. Estimation theory with applications to communications and control [M]. New York: McGraw-Hill Book Company, 1971.

[9] SORENSON, H W. Parameter estimation, principles and problems [M]. New York: Marcel Dekker, 1980.

第 4 章　基于状态估计的近距离相对导航

在第 3 章中，深入研究了基于标志点静态几何关系的航天器相对位姿确定方法。本章从状态估计的角度出发，将深入研究基于状态估计的近距离相对导航方法，结合航天器相对动力学设计导航滤波器，实现两航天器间相对位置、速度和姿态的估计。

首先，基于第 2 章提出的 VISNAV 观测模型 2，提出了一种新的基于 EKF 的航天器间相对位姿确定算法。与传统算法相比，该算法额外估计了两航天器体系相对于主星 LVLH 系的姿态，从而不必要去做主星本体系与 LVLH 系重合这一假设，其实质是不需要先验的主星绝对姿态信息。

其次，在前面的基础上，进一步提出基于 CKF 的航天器间相对位姿确定算法。与 EKF 相比，CKF 可以容许较大的初始条件误差，且不需要计算雅可比导数矩阵。利用传播的四元数容积点集的加权平均值作为参考四元数，进一步改进所提的 CKF 导航滤波器，在一定程度上提高滤波性能。

最后，考虑主星由于能料耗尽或系统故障在空间处于翻滚状态，在缺少主星惯性参数和角速度先验信息情形下，进一步推导主星失控翻滚情形下的相对位姿确定方法。

考虑到 C-W 方程是椭圆轨道相对运动动力学方程的特例，因此本章提出的方法同样适用于主星轨道为圆或近圆轨道情形。

4.1　基于 EKF 的航天器相对位姿确定方法

4.1.1　扩展卡尔曼滤波

在实际工程应用中，扩展卡尔曼滤波[1]常被用于处理非线性估计问题。考虑连续时间非线性系统：

$$\dot{\boldsymbol{x}}(t)=\boldsymbol{f}(\boldsymbol{x}(t),\boldsymbol{u}(t),t)+\boldsymbol{G}(t)\boldsymbol{w}(t) \tag{4.1}$$

$$\tilde{\boldsymbol{z}}(t)=\boldsymbol{h}(\boldsymbol{x}(t),t)+\boldsymbol{v}(t) \tag{4.2}$$

式中：$\boldsymbol{x}(t)\in\mathbb{R}^n$，$\boldsymbol{u}(t)\in\mathbb{R}^s$和$\tilde{\boldsymbol{z}}(t)\in\mathbb{R}^m$分别为 t 时刻状态矢量、输入矢量和观测矢量；$\boldsymbol{f}(\cdot)$和$\boldsymbol{h}(\cdot)$为连续可微的非线性函数；$\boldsymbol{G}(t)\in\mathbb{R}^{n\times l}$为过程噪声

分布矩阵；$\boldsymbol{w}(t)\in\mathbb{R}^{l}$和$\boldsymbol{v}(t)\in\mathbb{R}^{m}$分别为零均值高斯分布的过程噪声和观测噪声矢量，满足

$$\begin{cases}E\{\boldsymbol{w}(t)\boldsymbol{w}^{\mathrm{T}}(\tau)\}=\boldsymbol{Q}(t)\delta(t-\tau)\\E\{\boldsymbol{v}(t)\boldsymbol{v}^{\mathrm{T}}(\tau)\}=\boldsymbol{R}(t)\delta(t-\tau)\\E\{\boldsymbol{v}(t)\boldsymbol{w}^{\mathrm{T}}(\tau)\}=\boldsymbol{0}\end{cases}\tag{4.3}$$

将式（4.3）状态方程和观测方程在标称状态$\bar{\boldsymbol{x}}(t)$近傍泰勒展开，且取一阶项可得

$$\boldsymbol{f}(\boldsymbol{x}(t),\boldsymbol{u}(t),t)\cong\boldsymbol{f}(\bar{\boldsymbol{x}}(t),\boldsymbol{u}(t),t)+\left.\frac{\partial\boldsymbol{f}}{\partial\boldsymbol{x}}\right|_{\bar{\boldsymbol{x}}(t)}[\boldsymbol{x}(t)-\bar{\boldsymbol{x}}(t)]\tag{4.4}$$

$$\boldsymbol{h}(\boldsymbol{x}(t),t)\cong\boldsymbol{h}(\bar{\boldsymbol{x}}(t),t)+\left.\frac{\partial\boldsymbol{h}}{\partial\boldsymbol{x}}\right|_{\bar{\boldsymbol{x}}(t)}[\boldsymbol{x}(t)-\bar{\boldsymbol{x}}(t)]\tag{4.5}$$

在 EKF 中，当前状态估计值被用来当作标称状态，即$\bar{\boldsymbol{x}}(t)=\hat{\boldsymbol{x}}(t)$。此时，对式（4.4）和式（4.5）两边同时取期望，可得

$$\boldsymbol{E}\{\boldsymbol{f}(\boldsymbol{x}(t),\boldsymbol{u}(t),t)\}=\boldsymbol{f}(\hat{\boldsymbol{x}}(t),\boldsymbol{u}(t),t)\tag{4.6}$$

$$\boldsymbol{E}\{\boldsymbol{h}(\boldsymbol{x}(t),t)\}=\boldsymbol{h}(\hat{\boldsymbol{x}}(t),t)\tag{4.7}$$

因此，可得 EKF 状态估计结构为

$$\dot{\hat{\boldsymbol{x}}}(t)=\boldsymbol{f}(\hat{\boldsymbol{x}}(t),\boldsymbol{u}(t),t)+\boldsymbol{K}(t)[\tilde{\boldsymbol{z}}(t)-\boldsymbol{h}(\hat{\boldsymbol{x}}(t),t)]\tag{4.8}$$

定义状态估计误差为

$$\Delta\boldsymbol{x}\equiv\boldsymbol{x}-\hat{\boldsymbol{x}}\tag{4.9}$$

对式（4.9）求导，并代入式（4.1）和式（4.8），整理可得误差状态微分方程：

$$\Delta\dot{\boldsymbol{x}}(t)=[\boldsymbol{F}(\hat{\boldsymbol{x}}(t),t)-\boldsymbol{K}(t)\boldsymbol{H}(\hat{\boldsymbol{x}}(t),t)]\Delta\boldsymbol{x}(t)+\boldsymbol{G}(t)\boldsymbol{w}(t)-\boldsymbol{K}(t)\boldsymbol{v}(t)\tag{4.10}$$

其中，

$$\left.\frac{\partial\boldsymbol{f}}{\partial\boldsymbol{x}}\right|_{\hat{\boldsymbol{x}}(t)}=\boldsymbol{F}(\hat{\boldsymbol{x}}(t),t),\quad\left.\frac{\partial\boldsymbol{h}}{\partial\boldsymbol{x}}\right|_{\hat{\boldsymbol{x}}(t)}=\boldsymbol{H}(\hat{\boldsymbol{x}}(t),t)\tag{4.11}$$

由式（4.8）可以看出，增益$\boldsymbol{K}(t)$越大，则估计值越依赖观测量；而增益$\boldsymbol{K}(t)$越小，则估计值越依赖模型。从控制观点出发，滤波本质上是一种考虑观测量后的反馈结构，通过选择合适的增益矩阵$\boldsymbol{K}(t)$，使得$\boldsymbol{E}(t)=\boldsymbol{F}(\hat{\boldsymbol{x}}(t),t)-\boldsymbol{K}(t)\boldsymbol{H}(\hat{\boldsymbol{x}}(t),t)$稳定。在频率域中，滤波器的角频率（或带宽）是由$|\boldsymbol{E}(t)|$决定的。随着增益$\boldsymbol{K}(t)$增大，滤波器的角频率增大，估计值中会引入更多的高频噪声；相反，随着增益$\boldsymbol{K}(t)$减小，带宽减小，则进入滤波系统的噪声就减少，但如果$\boldsymbol{K}(t)$过小，那么估计误差收敛到零则非常慢。因此，需要合理选择增益$\boldsymbol{K}(t)$。对于时不变系统，误差方差阵$\boldsymbol{P}$很快收敛到一

个稳定值，因此可以利用稳定状态方差预先计算常值增益 $\boldsymbol{K}$，大大减少计算量，尽管从严格意义上来讲，这是一个次优方法。实际上，基于动力学模型和观测模型的随机过程，卡尔曼滤波器提供了一个科学严谨的配置估计器极点的理论方法。

工程实际问题中，经常会遇到状态方程是时间连续的，但观测方程是离散的情况，该滤波问题称为连续-离散滤波。表 4.1 给出了连续-离散扩展卡尔曼滤波方程。

表 4.1　连续-离散扩展卡尔曼滤波方程

模　型	$\dot{\boldsymbol{x}}(t)=\boldsymbol{f}(\boldsymbol{x}(t),\boldsymbol{u}(t),t)+\boldsymbol{G}(t)\boldsymbol{w}(t)$, $\boldsymbol{w}(t)\sim N(\boldsymbol{0},\boldsymbol{Q}(t))$ $\tilde{\boldsymbol{z}}_k=\boldsymbol{h}(\boldsymbol{x}_k)+\boldsymbol{v}_k$, $\boldsymbol{v}_k\sim N(\boldsymbol{0},\boldsymbol{R}_k)$
初始化	$\hat{\boldsymbol{x}}(t_0)=\hat{\boldsymbol{x}}_0$, $\boldsymbol{P}(t_0)=\boldsymbol{E}\{\Delta\boldsymbol{x}(t_0)\Delta\boldsymbol{x}^{\mathrm{T}}(t_0)\}$
增益及测量灵敏度矩阵	$\boldsymbol{K}_k=\boldsymbol{P}_k^-\boldsymbol{H}_k^{\mathrm{T}}(\hat{\boldsymbol{x}}_k^-)[\boldsymbol{H}_k(\hat{\boldsymbol{x}}_k^-)\boldsymbol{P}_k^-\boldsymbol{H}_k^{\mathrm{T}}(\hat{\boldsymbol{x}}_k^-)+\boldsymbol{R}_k]^{-1}$ $\boldsymbol{H}_k(\hat{\boldsymbol{x}}_k^-)\equiv\left.\dfrac{\partial\boldsymbol{h}}{\partial\boldsymbol{x}}\right\|_{\hat{\boldsymbol{x}}_k^-}$
更新	$\hat{\boldsymbol{x}}_k^+=\hat{\boldsymbol{x}}_k^-+\boldsymbol{K}_k[\tilde{\boldsymbol{z}}_k-\boldsymbol{h}(\hat{\boldsymbol{x}}_k^-)]$ $\boldsymbol{P}_k^+=[\boldsymbol{I}-\boldsymbol{K}_k\boldsymbol{H}_k(\hat{\boldsymbol{x}}_k^-)]\boldsymbol{P}_k^-$
传播	$\dot{\hat{\boldsymbol{x}}}(t)=\boldsymbol{f}(\hat{\boldsymbol{x}}(t),\boldsymbol{u}(t),t)$ $\hat{\boldsymbol{P}}(t)=\boldsymbol{F}(\hat{\boldsymbol{x}}(t),t)\boldsymbol{P}(t)+\boldsymbol{P}(t)\boldsymbol{F}^{\mathrm{T}}(\hat{\boldsymbol{x}}(t),t)+\boldsymbol{G}(t)\boldsymbol{Q}(t)\boldsymbol{G}^{\mathrm{T}}(t)$ $\boldsymbol{F}(\hat{\boldsymbol{x}}(t),t)\equiv\left.\dfrac{\partial\boldsymbol{h}}{\partial\boldsymbol{x}}\right\|_{\hat{\boldsymbol{x}}(t)}$

当滤波器采用误差状态量作为滤波变量时，滤波更新方程为

$$\Delta\hat{\boldsymbol{x}}_k^+=\boldsymbol{K}_k[\tilde{\boldsymbol{z}}_k-\boldsymbol{h}(\hat{\boldsymbol{x}}_k^-)] \tag{4.12}$$

每次滤波更新后，滤波器都必须重置，即令 $\Delta\hat{\boldsymbol{x}}_k^+=\boldsymbol{0}$。

当遇到大维数观测量时，传统的卡尔曼滤波形式需要计算与 $\boldsymbol{R}_k$ 同阶数的逆矩阵，给数值计算带来一定困难。目前有两种处理方法：①采用信息滤波技术，其在计算增益 $\boldsymbol{K}_k$ 求逆时，始终只用执行与状态维数大小一致的逆操作；②采用观测量连续处理方法，将同一时刻观测量逐个处理。

4.1.2　EKF 导航滤波器设计

4.1.2.1　系统状态方程及其线性化

根据第 2 章中描述的航天器相对动力学与敏感器模型，可得状态估计方程

$$\dot{\hat{\boldsymbol{q}}}_{d/H}=\frac{1}{2}\boldsymbol{\Xi}(\hat{\boldsymbol{q}}_{d/H})\hat{\boldsymbol{\omega}}_{d/H}^{d} \tag{4.13}$$

$$\hat{\boldsymbol{\omega}}_{d/H}^{d}=\hat{\boldsymbol{\omega}}_{d/I}^{d}-\boldsymbol{A}_{H}^{d}(\hat{\boldsymbol{q}}_{d/H})\hat{\boldsymbol{\omega}}_{H/I}^{H} \tag{4.14}$$

$$\dot{\hat{\boldsymbol{q}}}_{c/H}=\frac{1}{2}\boldsymbol{\Xi}(\hat{\boldsymbol{q}}_{c/H})\hat{\boldsymbol{\omega}}_{c/H}^{c} \tag{4.15}$$

$$\hat{\boldsymbol{\omega}}_{c/H}^{c}=\hat{\boldsymbol{\omega}}_{c/I}^{c}-\boldsymbol{A}_{H}^{c}(\hat{\boldsymbol{q}}_{c/H})\hat{\boldsymbol{\omega}}_{H/I}^{H} \tag{4.16}$$

$$\hat{\boldsymbol{\omega}}_{d/I}^{d}=\widetilde{\boldsymbol{\omega}}_{d/I}^{d}-\hat{\boldsymbol{\beta}}_{d} \tag{4.17}$$

$$\hat{\boldsymbol{\omega}}_{c/I}^{c}=\widetilde{\boldsymbol{\omega}}_{c/I}^{c}-\hat{\boldsymbol{\beta}}_{c} \tag{4.18}$$

$$\dot{\hat{\boldsymbol{x}}}_{p}=\boldsymbol{f}(\hat{\boldsymbol{x}}_{p}) \tag{4.19}$$

$$\hat{\boldsymbol{\omega}}_{H/I}^{H}=[0\quad 0\quad \hat{\dot{\theta}}]^{\mathrm{T}} \tag{4.20}$$

$$\dot{\hat{\boldsymbol{\beta}}}_{d}=\boldsymbol{0} \tag{4.21}$$

$$\dot{\hat{\boldsymbol{\beta}}}_{c}=\boldsymbol{0} \tag{4.22}$$

针对四元数的正交性约束会造成误差方差阵奇异这一问题，本节采用乘性扩展卡尔曼滤波（Multiplicative Extended Kalman Filter，MEKF）技术[2]，即通过估计无约束的三分量姿态误差参数并利用四元数乘法为航天器提供全局非奇异姿态描述。因此，定义状态变量

$$\begin{aligned}\boldsymbol{x}&\equiv[\boldsymbol{q}_{d/H}^{\mathrm{T}}\quad \boldsymbol{q}_{c/H}^{\mathrm{T}}\quad \boldsymbol{\beta}_{d}^{\mathrm{T}}\quad \boldsymbol{\beta}_{c}^{\mathrm{T}}\quad \boldsymbol{x}_{p}^{\mathrm{T}}]^{\mathrm{T}}\\&=[\boldsymbol{q}_{d/H}^{\mathrm{T}}\quad \boldsymbol{q}_{c/H}^{\mathrm{T}}\quad \boldsymbol{\beta}_{d}^{\mathrm{T}}\quad \boldsymbol{\beta}_{c}^{\mathrm{T}}\quad \boldsymbol{\rho}^{\mathrm{T}}\quad \dot{\boldsymbol{\rho}}^{\mathrm{T}}\quad r_{c}\quad \dot{r}_{c}\quad \theta\quad \dot{\theta}]^{\mathrm{T}}\end{aligned} \tag{4.23}$$

和误差状态变量

$$\Delta\boldsymbol{x}\equiv[\delta\boldsymbol{\alpha}_{d/H}^{\mathrm{T}}\quad \delta\boldsymbol{\alpha}_{c/H}^{\mathrm{T}}\quad \Delta\boldsymbol{\beta}_{d}^{\mathrm{T}}\quad \Delta\boldsymbol{\beta}_{c}^{\mathrm{T}}\quad \Delta\boldsymbol{\rho}^{\mathrm{T}}\quad \Delta\dot{\boldsymbol{\rho}}^{\mathrm{T}}\quad \Delta r_{c}\quad \Delta\dot{r}_{c}\quad \Delta\theta\quad \Delta\dot{\theta}]^{\mathrm{T}} \tag{4.24}$$

其中，$\delta\boldsymbol{\alpha}$ 为航天器体坐标系下定义的姿态误差矢量（滚转、俯仰和偏航误差）。对应的姿态误差阵可表示为

$$\boldsymbol{A}(\delta\boldsymbol{q})\approx\boldsymbol{I}_{3\times3}-[\delta\boldsymbol{\alpha}\times] \tag{4.25}$$

式（4.24）中的其他误差项 $\Delta\bullet$ 则定义为 $\Delta\bullet\equiv\bullet-\hat{\bullet}$。

线性化可得到姿态误差矢量 $\delta\boldsymbol{\alpha}_{d/H}$ 的动力学模型

$$\delta\dot{\boldsymbol{\alpha}}_{d/H}=-[\hat{\boldsymbol{\omega}}_{d/I}^{d}\times]\delta\boldsymbol{\alpha}_{d/H}+\Delta\boldsymbol{\omega}_{d/I}^{d}-\boldsymbol{A}_{H}^{d}(\hat{\boldsymbol{q}}_{d/H})\Delta\boldsymbol{\omega}_{H/I}^{H} \tag{4.26}$$

其中，$\Delta\boldsymbol{\omega}_{d/I}^{d}=\boldsymbol{\omega}_{d/I}^{d}-\hat{\boldsymbol{\omega}}_{d/I}^{d}$，$\Delta\boldsymbol{\omega}_{H/I}^{H}=\boldsymbol{\omega}_{H/I}^{H}-\hat{\boldsymbol{\omega}}_{H/I}^{H}$。误差四元数的第四个分量的导数为零。进一步，将 $\Delta\boldsymbol{\omega}_{d/I}^{d}=-(\Delta\boldsymbol{\beta}_{d}+\boldsymbol{\eta}_{dv})$ 和 $\Delta\boldsymbol{\omega}_{H/I}^{H}=\boldsymbol{n}\cdot\Delta\dot{\theta}$ 代入式（4.26）可推导得到

$$\delta\dot{\boldsymbol{\alpha}}_{d/H}=-[\hat{\boldsymbol{\omega}}_{d/I}^{d}\times]\delta\boldsymbol{\alpha}_{d/H}-\Delta\boldsymbol{\beta}_{d}-\boldsymbol{A}_{H}^{d}(\hat{\boldsymbol{q}}_{d/H})\boldsymbol{n}\cdot\Delta\dot{\theta}-\boldsymbol{\eta}_{dv} \tag{4.27}$$

其中，$\Delta\boldsymbol{\beta}_d \equiv \boldsymbol{\beta}_d - \hat{\boldsymbol{\beta}}_d$，$\Delta\dot{\theta} \equiv \dot{\theta} - \hat{\dot{\theta}}$，$\boldsymbol{n} = [0\quad 0\quad 1]^{\mathrm{T}}$。

同样地，推导可得 $\delta\boldsymbol{\alpha}_{c/H}$ 的线性化动力学模型为

$$\delta\dot{\boldsymbol{\alpha}}_{c/H} = -[\hat{\boldsymbol{\omega}}^c_{c/I}\times]\delta\boldsymbol{\alpha}_{c/H} - \Delta\boldsymbol{\beta}_c - \boldsymbol{A}^c_H(\hat{\boldsymbol{q}}_{c/H})\boldsymbol{n}\cdot\Delta\dot{\theta} - \boldsymbol{\eta}_{cv} \tag{4.28}$$

过程噪声矢量 $\boldsymbol{w}$ 及其谱密度矩阵 $\boldsymbol{Q}$ 分别定义为

$$\boldsymbol{w} \equiv [\boldsymbol{\eta}^{\mathrm{T}}_{dv}\quad \boldsymbol{\eta}^{\mathrm{T}}_{cv}\quad \boldsymbol{\eta}^{\mathrm{T}}_{du}\quad \boldsymbol{\eta}^{\mathrm{T}}_{cu}\quad \boldsymbol{\varpi}^{\mathrm{T}}]^{\mathrm{T}} \tag{4.29}$$

$$\boldsymbol{Q} = \begin{bmatrix} \sigma^2_{dv}\boldsymbol{I}_{3\times3} & \boldsymbol{0}_{3\times3} & \boldsymbol{0}_{3\times3} & \boldsymbol{0}_{3\times3} & \boldsymbol{0}_{3\times3} \\ \boldsymbol{0}_{3\times3} & \sigma^2_{cv}\boldsymbol{I}_{3\times3} & \boldsymbol{0}_{3\times3} & \boldsymbol{0}_{3\times3} & \boldsymbol{0}_{3\times3} \\ \boldsymbol{0}_{3\times3} & \boldsymbol{0}_{3\times3} & \sigma^2_{du}\boldsymbol{I}_{3\times3} & \boldsymbol{0}_{3\times3} & \boldsymbol{0}_{3\times3} \\ \boldsymbol{0}_{3\times3} & \boldsymbol{0}_{3\times3} & \boldsymbol{0}_{3\times3} & \sigma^2_{cu}\boldsymbol{I}_{3\times3} & \boldsymbol{0}_{3\times3} \\ \boldsymbol{0}_{3\times3} & \boldsymbol{0}_{3\times3} & \boldsymbol{0}_{3\times3} & \boldsymbol{0}_{3\times3} & \sigma^2_{\varpi}\boldsymbol{I}_{3\times3} \end{bmatrix} \tag{4.30}$$

因此，误差状态方程可写为

$$\Delta\dot{\boldsymbol{x}} = \boldsymbol{F}\Delta\boldsymbol{x} + \boldsymbol{G}\boldsymbol{w} \tag{4.31}$$

其中，

$$\boldsymbol{F} \equiv \begin{bmatrix} -[\hat{\boldsymbol{\omega}}^d_{d/I}\times] & \boldsymbol{0}_{3\times3} & -\boldsymbol{I}_{3\times3} & \boldsymbol{0}_{3\times3} & \boldsymbol{0}_{3\times9} & -\boldsymbol{A}^d_H(\hat{\boldsymbol{q}}_{d/H})\boldsymbol{n} \\ \boldsymbol{0}_{3\times3} & -[\hat{\boldsymbol{\omega}}^c_{c/I}\times] & \boldsymbol{0}_{3\times3} & -\boldsymbol{I}_{3\times3} & \boldsymbol{0}_{3\times9} & -\boldsymbol{A}^c_H(\hat{\boldsymbol{q}}_{c/H})\boldsymbol{n} \\ \boldsymbol{0}_{3\times3} & \boldsymbol{0}_{3\times3} & \boldsymbol{0}_{3\times3} & \boldsymbol{0}_{3\times3} & \boldsymbol{0}_{3\times10} & \\ \boldsymbol{0}_{3\times3} & \boldsymbol{0}_{3\times3} & \boldsymbol{0}_{3\times3} & \boldsymbol{0}_{3\times3} & \boldsymbol{0}_{3\times10} & \\ \boldsymbol{0}_{10\times3} & \boldsymbol{0}_{10\times3} & \boldsymbol{0}_{10\times3} & \boldsymbol{0}_{10\times3} & \left.\dfrac{\partial \boldsymbol{f}(\boldsymbol{x}_p)}{\partial \boldsymbol{x}_p}\right|_{\hat{\boldsymbol{x}}_p} & \end{bmatrix},\quad \boldsymbol{n} = [0\quad 0\quad 1]^{\mathrm{T}} \tag{4.32}$$

$$\boldsymbol{G} \equiv \begin{bmatrix} -\boldsymbol{I}_{3\times3} & \boldsymbol{0}_{3\times3} & \boldsymbol{0}_{3\times3} & \boldsymbol{0}_{3\times3} & \boldsymbol{0}_{3\times3} \\ \boldsymbol{0}_{3\times3} & -\boldsymbol{I}_{3\times3} & \boldsymbol{0}_{3\times3} & \boldsymbol{0}_{3\times3} & \boldsymbol{0}_{3\times3} \\ \boldsymbol{0}_{3\times3} & \boldsymbol{0}_{3\times3} & \boldsymbol{I}_{3\times3} & \boldsymbol{0}_{3\times3} & \boldsymbol{0}_{3\times3} \\ \boldsymbol{0}_{3\times3} & \boldsymbol{0}_{3\times3} & \boldsymbol{0}_{3\times3} & \boldsymbol{I}_{3\times3} & \boldsymbol{0}_{3\times3} \\ \boldsymbol{0}_{3\times3} & \boldsymbol{0}_{3\times3} & \boldsymbol{0}_{3\times3} & \boldsymbol{0}_{3\times3} & \boldsymbol{0}_{3\times3} \\ \boldsymbol{0}_{3\times3} & \boldsymbol{0}_{3\times3} & \boldsymbol{0}_{3\times3} & \boldsymbol{0}_{3\times3} & \boldsymbol{I}_{3\times3} \\ \boldsymbol{0}_{2\times3} & \boldsymbol{0}_{2\times3} & \boldsymbol{0}_{2\times3} & \boldsymbol{0}_{2\times3} & \boldsymbol{0}_{2\times3} \\ \boldsymbol{0}_{2\times3} & \boldsymbol{0}_{2\times3} & \boldsymbol{0}_{2\times3} & \boldsymbol{0}_{2\times3} & \boldsymbol{0}_{2\times3} \end{bmatrix} \tag{4.33}$$

式（4.32）中推导的偏导数矩阵 $\partial\boldsymbol{f}(\boldsymbol{x}_p)/\partial\boldsymbol{x}_p$ 详见附录 B。

4.1.2.2　状态和方差传播

状态变量 $\hat{\boldsymbol{x}}$ 可通过积分方程（4.13）~方程（4.22）得到。方差传播的离

散形式可以表示成

$$\boldsymbol{P}_{k+1}^{-}=\boldsymbol{\Phi}_k\boldsymbol{P}_k^{+}\boldsymbol{\Phi}_k^{\mathrm{T}}+\boldsymbol{Q}_k \tag{4.34}$$

式中：$\boldsymbol{\Phi}_k$ 为状态转移阵；$\boldsymbol{Q}_k$ 为离散的过程噪声方差阵。Loan 给出了这些矩阵的数值解[3]，首先取时间间隔 Δt，构造矩阵

$$\boldsymbol{\mathcal{A}}=\begin{bmatrix}-\boldsymbol{F} & \boldsymbol{GQG}^{\mathrm{T}}\\ \boldsymbol{0} & \boldsymbol{F}\end{bmatrix}\Delta t \tag{4.35}$$

其中，矩阵 $\boldsymbol{F}$，$\boldsymbol{G}$，$\boldsymbol{Q}$ 均已定义。然后，计算方程（4.35）的矩阵指数

$$\boldsymbol{\mathcal{B}}=e^{\boldsymbol{\mathcal{A}}}\equiv\begin{bmatrix}\boldsymbol{\mathcal{B}}_{11} & \boldsymbol{\mathcal{B}}_{12}\\ \boldsymbol{0} & \boldsymbol{\mathcal{B}}_{22}\end{bmatrix}=\begin{bmatrix}\boldsymbol{\mathcal{B}}_{11} & \boldsymbol{\Phi}_k^{-1}\boldsymbol{Q}_k\\ \boldsymbol{0} & \boldsymbol{\Phi}_k^{\mathrm{T}}\end{bmatrix} \tag{4.36}$$

于是可得

$$\boldsymbol{\Phi}_k=\boldsymbol{\mathcal{B}}_{22}^{\mathrm{T}} \tag{4.37}$$

$$\boldsymbol{Q}_k=\boldsymbol{\Phi}_k\boldsymbol{\mathcal{B}}_{12} \tag{4.38}$$

不过，上述方法需要计算矩阵指数，在 MATLAB 环境下较容易实现，但在 C 语言环境中仍需要编写矩阵指数函数。参考文献［4-5］中运用矩阵论中二次型原理推导了连续系统离散化的一般公式，指出离散过程噪声协方差阵具有形式

$$\boldsymbol{Q}_k=\sum_{J=1}^{L}\sum_{I=1}^{L}\frac{1}{(I+J-1)}\frac{1}{(I-1)!}\frac{1}{(J-1)!}\boldsymbol{F}^{I-1}(t_k)\boldsymbol{G}(t_k)\boldsymbol{Q}(t_k)\boldsymbol{G}^{\mathrm{T}}(t_k)\left(\boldsymbol{F}^{\mathrm{T}}(t_k)\right)^{J-1}\Delta t^{I+J-1} \tag{4.39}$$

可根据精度要求选取适当的 L，一般选取 $L=3\sim4$。

值得注意的是，当采样周期 Δt 足够短时，式（4.38）中的离散过程噪声方差阵可由 $\Delta t\boldsymbol{GQG}^{\mathrm{T}}$ 近似。

4.1.2.3 观测方程及其线性化

经过一系列简单的数学推导，可得到 VISNAV 测量灵敏度矩阵

$$\boldsymbol{H}_k(\hat{\boldsymbol{x}}_k^{-})=\begin{bmatrix}[\boldsymbol{A}_H^d(\hat{\boldsymbol{q}}_{d/H}^{-})\hat{\boldsymbol{r}}_1^{-}\times] & \dfrac{\partial\hat{\boldsymbol{b}}_1^{-}}{\partial\delta\boldsymbol{\alpha}_{c/H}} & \boldsymbol{0}_{3\times3} & \boldsymbol{0}_{3\times3} & \dfrac{\partial\hat{\boldsymbol{b}}_1^{-}}{\partial\hat{\boldsymbol{\rho}}^{-}} & \boldsymbol{0}_{3\times7}\\ \vdots & \vdots & \vdots & \vdots & \vdots & \vdots\\ [\boldsymbol{A}_H^d(\hat{\boldsymbol{q}}_{d/H}^{-})\hat{\boldsymbol{r}}_N^{-}\times] & \dfrac{\partial\hat{\boldsymbol{b}}_N^{-}}{\partial\delta\boldsymbol{\alpha}_{c/H}} & \boldsymbol{0}_{3\times3} & \boldsymbol{0}_{3\times3} & \dfrac{\partial\hat{\boldsymbol{b}}_N^{-}}{\partial\hat{\boldsymbol{\rho}}^{-}} & \boldsymbol{0}_{3\times7}\end{bmatrix} \tag{4.40}$$

式中：$\hat{\boldsymbol{r}}_i^{-}$ 可由式（2.40）在 $\hat{\boldsymbol{\rho}}^{-}\equiv[\hat{x}^{-}\quad\hat{y}^{-}\quad\hat{z}^{-}]^{\mathrm{T}}$ 和 $[\boldsymbol{A}_H^c(\hat{\boldsymbol{q}}_{c/H}^{-})]^{\mathrm{T}}$ 处计算得到。

偏导数矩阵 $\partial\hat{\boldsymbol{b}}_i^{-}/\partial\hat{\boldsymbol{\rho}}^{-}$ 可表示为

$$\frac{\partial \hat{\boldsymbol{b}}_i^-}{\partial \hat{\boldsymbol{\rho}}^-}=\boldsymbol{A}_H^d(\hat{\boldsymbol{q}}_{d/H}^-)\frac{\partial \hat{\boldsymbol{r}}_i^-}{\partial \hat{\boldsymbol{\rho}}^-} \tag{4.41}$$

其中，

$$\frac{\partial \hat{\boldsymbol{r}}_i^-}{\partial \hat{\boldsymbol{\rho}}^-}=\frac{1}{\hat{s}_i^-}\begin{bmatrix} -[(\hat{Y}_i'^- -\hat{y}^-)^2+(\hat{Z}_i'^- -\hat{z}^-)^2 & (\hat{X}_i'^- -\hat{x}^-)(\hat{Y}_i'^- -\hat{y}^-) & (\hat{X}_i'^- -\hat{x}^-)(\hat{Z}_i'^- -\hat{z}^-) \\ (\hat{X}_i'^- -\hat{x}^-)(\hat{Y}_i'^- -\hat{y}^-) & -[(\hat{X}_i'^- -\hat{x}^-)^2+(Z_i'-\hat{z}^-)^2] & (\hat{Y}_i'^- -\hat{y}^-)(\hat{Z}_i'^- -\hat{z}^-) \\ (\hat{X}_i'^- -\hat{x}^-)(\hat{Z}_i'^- -\hat{z}^-) & (\hat{Y}_i'^- -\hat{y}^-)(\hat{Z}_i'^- -\hat{z}^-) & -[(\hat{X}_i'^- -\hat{x}^-)^2+(\hat{Y}_i'^- -\hat{y}^-)^2] \end{bmatrix} \tag{4.42}$$

式中：$\hat{s}_i^- \equiv [(X_i'-\hat{x}^-)^2+(Y_i'-\hat{y}^-)^2+(Z'_i-\hat{z}^-)^2]^{3/2}$；$\hat{\boldsymbol{\mathcal{X}}}_i'^- =[\boldsymbol{A}_H^c(\hat{\boldsymbol{q}}_{c/H}^-)]^{\mathrm{T}}\boldsymbol{\mathcal{X}}_i$。

另外，根据偏微分的链式法则，推导可得式（4.40）中的偏导数矩阵 $\partial \hat{\boldsymbol{b}}_i^-/\partial\delta\boldsymbol{\alpha}_{c/H}$ 为[6]

$$\frac{\partial \hat{\boldsymbol{b}}_i^-}{\partial\, \delta\boldsymbol{\alpha}_{c/H}}=\frac{\partial \hat{\boldsymbol{b}}_i^-}{\partial \hat{\boldsymbol{\mathcal{X}}}_i'^-}\frac{\partial \hat{\boldsymbol{\mathcal{X}}}_i'^-}{\partial\, \delta\boldsymbol{\alpha}_{c/H}} \tag{4.43}$$

$$\frac{\partial \hat{\boldsymbol{b}}_i^-}{\partial \hat{\boldsymbol{\mathcal{X}}}_i'^-}=-\boldsymbol{A}_H^d(\hat{\boldsymbol{q}}_{d/H}^-)\frac{\partial \hat{\boldsymbol{r}}_i^-}{\partial \hat{\boldsymbol{\rho}}^-} \tag{4.44}$$

利用式（2.41）和式（4.25），可得 $\boldsymbol{\mathcal{X}}_i'$ 和 $\delta\boldsymbol{\alpha}_{c/H}$ 关系为

$$\begin{aligned}\boldsymbol{\mathcal{X}}_i' &=[(\boldsymbol{I}_{3\times3}-[\delta\boldsymbol{\alpha}_{c/H}\times])\boldsymbol{A}_H^c(\hat{\boldsymbol{q}}_{c/H}^-)]^{\mathrm{T}}\boldsymbol{\mathcal{X}}_i \\ &=[\boldsymbol{A}_H^c(\hat{\boldsymbol{q}}_{c/H}^-)]^{\mathrm{T}}(\boldsymbol{I}_{3\times3}+[\delta\boldsymbol{\alpha}_{c/H}\times])\boldsymbol{\mathcal{X}}_i\end{aligned} \tag{4.45}$$

因此，可得偏导数矩阵 $\partial\boldsymbol{\mathcal{X}}_i'^-/\partial\delta\boldsymbol{\alpha}_{c/H}$ 为

$$\frac{\partial \boldsymbol{\mathcal{X}}_i'^-}{\partial\, \delta\boldsymbol{\alpha}_{c/H}}=-[\boldsymbol{A}_H^c(\hat{\boldsymbol{q}}_{c/H}^-)]^{\mathrm{T}}[\boldsymbol{\mathcal{X}}_i\times] \tag{4.46}$$

联合式（4.43）~式（4.46），可得

$$\frac{\partial \hat{\boldsymbol{b}}_i^-}{\partial\, \delta\boldsymbol{\alpha}_{c/H}}=\boldsymbol{A}_H^d(\hat{\boldsymbol{q}}_{d/H}^-)\frac{\partial \hat{\boldsymbol{r}}_i^-}{\partial \hat{\boldsymbol{\rho}}^-}[\boldsymbol{A}_H^c(\hat{\boldsymbol{q}}_{c/H}^-)]^{\mathrm{T}}[\boldsymbol{\mathcal{X}}_i\times] \tag{4.47}$$

其中，$\partial \hat{\boldsymbol{r}}_i^-/\partial \hat{\boldsymbol{\rho}}^-$ 可由式（4.42）计算得到。表 4.2 概括了上述基于 EKF 的相对位姿确定算法，其中，矩阵 $\boldsymbol{P}$ 为误差状态方差阵，$\tilde{z}=[\tilde{\boldsymbol{b}}_1^{\mathrm{T}} \quad \tilde{\boldsymbol{b}}_2^{\mathrm{T}} \quad \cdots \quad \tilde{\boldsymbol{b}}_N^{\mathrm{T}}]^{\mathrm{T}}$ 为 VISNAV 系统的观测矢量。相对四元数 $\boldsymbol{q}_{d/H}$ 和 $\boldsymbol{q}_{c/H}$ 在每一步更新完成后需要

重新归一化。

表 4.2　基于 EKF 的相对位姿确定算法

初　始　化	$\hat{\boldsymbol{x}}(t_0)=\hat{\boldsymbol{x}}_0,\ \boldsymbol{P}(t_0)=\boldsymbol{P}_0$
增益及 测量灵敏度矩阵	$\boldsymbol{K}_k=\boldsymbol{P}_k^-\boldsymbol{H}_k^{\mathrm{T}}(\hat{\boldsymbol{x}}_k^-)[\boldsymbol{H}_k(\hat{\boldsymbol{x}}_k^-)\boldsymbol{P}_k^-\boldsymbol{H}_k^{\mathrm{T}}(\hat{\boldsymbol{x}}_k^-)+\boldsymbol{R}_k]^{-1}$ $\boldsymbol{H}_k(\hat{\boldsymbol{x}}_k^-)=\begin{bmatrix}[\boldsymbol{A}_H^d(\hat{\boldsymbol{q}}_{d/H}^-)\hat{\boldsymbol{r}}_1^-\times] & \dfrac{\partial\hat{\boldsymbol{b}}_1^-}{\partial\delta\boldsymbol{\alpha}_{c/H}} & \boldsymbol{0}_{3\times3} & \boldsymbol{0}_{3\times3} & \dfrac{\partial\hat{\boldsymbol{b}}_1^-}{\partial\hat{\boldsymbol{\rho}}^-} & \boldsymbol{0}_{3\times7}\\ \vdots & \vdots & \vdots & \vdots & \vdots & \vdots\\ [\boldsymbol{A}_H^d(\hat{\boldsymbol{q}}_{d/H}^-)\hat{\boldsymbol{r}}_N^-\times] & \dfrac{\partial\hat{\boldsymbol{b}}_N^-}{\partial\delta\boldsymbol{\alpha}_{c/H}} & \boldsymbol{0}_{3\times3} & \boldsymbol{0}_{3\times3} & \dfrac{\partial\hat{\boldsymbol{b}}_N^-}{\partial\hat{\boldsymbol{\rho}}^-} & \boldsymbol{0}_{3\times7}\end{bmatrix}$
更新	$\boldsymbol{P}_k^+=[\boldsymbol{I}-\boldsymbol{K}_k\boldsymbol{H}_k(\hat{\boldsymbol{x}}_k^-)]\boldsymbol{P}_k^-$ $\Delta\boldsymbol{x}_k^+=\boldsymbol{K}_k[\tilde{\boldsymbol{z}}_k-\boldsymbol{h}(\hat{\boldsymbol{x}}_k^-)]$ $\Delta\hat{\boldsymbol{x}}_k^+\equiv[\delta\hat{\boldsymbol{\alpha}}_{d/H_k}^{+\mathrm{T}}\quad\delta\hat{\boldsymbol{\alpha}}_{c/H_k}^{+\mathrm{T}}\quad\Delta\hat{\boldsymbol{\beta}}_{d_k}^{+\mathrm{T}}\quad\Delta\hat{\boldsymbol{\beta}}_{c_k}^{+\mathrm{T}}\quad\Delta\hat{\boldsymbol{x}}_{p_k}^{+\mathrm{T}}]^{\mathrm{T}}$ $\boldsymbol{h}_k(\hat{\boldsymbol{x}}_k^-)=\left.\begin{bmatrix}\boldsymbol{A}_H^d(\hat{\boldsymbol{q}}_{d/H}^-)\hat{\boldsymbol{r}}_1^-\\ \boldsymbol{A}_H^d(\hat{\boldsymbol{q}}_{d/H}^-)\hat{\boldsymbol{r}}_2^-\\ \vdots\\ \boldsymbol{A}_H^d(\hat{\boldsymbol{q}}_{d/H}^-)\hat{\boldsymbol{r}}_N^-\end{bmatrix}\right\rvert_{t_k}$ $\hat{\boldsymbol{q}}_{d/H_k}^+=\hat{\boldsymbol{q}}_{d/H_k}^-+\dfrac{1}{2}\boldsymbol{\Xi}(\hat{\boldsymbol{q}}_{d/H_k}^-)\delta\hat{\boldsymbol{\alpha}}_{d/H_k}^+$ $\hat{\boldsymbol{q}}_{c/H_k}^+=\hat{\boldsymbol{q}}_{c/H_k}^-+\dfrac{1}{2}\boldsymbol{\Xi}(\hat{\boldsymbol{q}}_{c/H_k}^-)\delta\hat{\boldsymbol{\alpha}}_{c/H_k}^+$ $\hat{\boldsymbol{\beta}}_{d_k}^+=\hat{\boldsymbol{\beta}}_{d_k}^-+\Delta\hat{\boldsymbol{\beta}}_{d_k}^+$ $\hat{\boldsymbol{\beta}}_{c_k}^+=\hat{\boldsymbol{\beta}}_{c_k}^-+\Delta\hat{\boldsymbol{\beta}}_{c_k}^+$ $\hat{\boldsymbol{x}}_{p_k}^+=\hat{\boldsymbol{x}}_{p_k}^-+\Delta\hat{\boldsymbol{x}}_{p_k}^+$
传播	$\hat{\boldsymbol{\omega}}_{d/I_k}^{d+}=\tilde{\boldsymbol{\omega}}_{d/I_k}^{d}-\hat{\boldsymbol{\beta}}_{d_k}^+$ $\hat{\boldsymbol{\omega}}_{c/I_k}^{c+}=\tilde{\boldsymbol{\omega}}_{c/I_k}^{c}-\hat{\boldsymbol{\beta}}_{c_k}^+$ $\hat{\boldsymbol{\omega}}_{H/I_k}^{H+}=[0\quad 0\quad \hat{\dot{\theta}}_k^+]^{\mathrm{T}}$ $\hat{\boldsymbol{q}}_{d/H_{k+1}}^-=\bar{\boldsymbol{\Omega}}(\hat{\boldsymbol{\omega}}_{d/I_k}^{d+})\bar{\boldsymbol{\Gamma}}(\hat{\boldsymbol{\omega}}_{H/I_k}^{H+})\hat{\boldsymbol{q}}_{d/H_k}^-$ $\hat{\boldsymbol{q}}_{c/H_{k+1}}^-=\bar{\boldsymbol{\Omega}}(\hat{\boldsymbol{\omega}}_{c/I_k}^{c+})\bar{\boldsymbol{\Gamma}}(\hat{\boldsymbol{\omega}}_{H/I_k}^{H+})\hat{\boldsymbol{q}}_{c/H_k}^-$ $\dot{\hat{\boldsymbol{x}}}_p=f(\hat{\boldsymbol{x}}_p)$ $\boldsymbol{P}_{k+1}^-=\boldsymbol{\Phi}_k\boldsymbol{P}_k^+\boldsymbol{\Phi}_k^{\mathrm{T}}+\boldsymbol{Q}_k$

4.1.3　仿真实验与分析

本节对所设计的 EKF 导航滤波器进行数值仿真验证。仿真参数设置与参考文献［7］基本一致，具体参数真值如表 4.3 所示。表 4.4 列出了主星上 6 个标志点的位置坐标。敏感器采样时间均为 1s，仿真时间为 300min。

表 4.3　仿真参数

仿真参数	数　值
初始陀螺漂移	$\boldsymbol{\beta}_d=[1,1,1]^{\mathrm{T}}((°)/\mathrm{h}),\boldsymbol{\beta}_c=[1,1,1]^{\mathrm{T}}((°)/\mathrm{h})$
陀螺噪声	$\sigma_{dv}=\sigma_{cv}=\sqrt{10}\times10^{-5}\mathrm{rad/s^{1/2}}$ $\sigma_{du}=\sigma_{cu}=\sqrt{10}\times10^{-10}\mathrm{rad/s^{3/2}}$
VISNAV 测量噪声	$\sigma_i=1.8'',i=1,2,\cdots,6$
扰动加速度过程噪声	$\sigma_{\varpi}=\sqrt{10}\times10^{-11}(\mathrm{m/s})^{3/2}$
角速度	$\boldsymbol{\omega}_{d/I}^{d}=[-0.002,0,0.0011]^{\mathrm{T}}\mathrm{rad/s},\boldsymbol{\omega}_{c/I}^{c}=[0,0.0011,-0.0011]^{\mathrm{T}}\mathrm{rad/s}$
初始四元数	$\boldsymbol{q}_{d/H}=[0,0,0,1]^{\mathrm{T}},\boldsymbol{q}_{c/H}=[0,0,0,1]^{\mathrm{T}}$
初始相对位置	$\boldsymbol{\rho}(0)=[200,200,100]^{\mathrm{T}}\mathrm{m}$
初始相对速度	$\dot{\boldsymbol{\rho}}(0)=[0.01,-0.4325,0.01]^{\mathrm{T}}\mathrm{m/s}$
初始主星轨道半径和径向速率	$r_c(0)=6.986417\times10^6\mathrm{m},\ \dot{r}_c(0)=0$
初始真近点角和真近点角速率	$\theta(0)=0,\ \dot{\theta}(0)=0.00108\mathrm{rad/s}$

表 4.4　标志点位置

序　号	X/m	Y/m	Z/m
1	0.5	0.5	0.0
2	-0.5	-0.5	0.0
3	-0.5	0.5	0.0
4	0.5	-0.5	0.0
5	0.2	0.5	0.1
6	0.0	0.2	-0.1

除了陀螺漂移初值设为零，其余状态量的滤波初值均设为真值。对于相对四元数 $\boldsymbol{q}_{d/H}$ 和 $\boldsymbol{q}_{c/H}$，初始姿态方差设为 $\boldsymbol{I}_{3\times3}(°)^2$；从星和主星初始陀螺漂移方差均为 $4\boldsymbol{I}_{3\times3}((°)/\mathrm{h})^2$；初始相对位置和速度方差分别为 $5\boldsymbol{I}_{3\times3}\mathrm{m}^2$ 和 $0.02\boldsymbol{I}_{3\times3}(\mathrm{m/s})^2$；主星初始轨道半径和径向速率方差分别为 $1000\mathrm{m}^2$ 和 $0.01(\mathrm{m/s})^2$；初始真近点角和真近点角速率方差分别为 $1\times10^{-4}\mathrm{rad}^2$ 和 $1\times10^{-4}(\mathrm{rad/s})^2$。因此，滤波初始方差阵为

$$\boldsymbol{P}_0=\mathrm{blkdiag}[(\pi/180)^2\boldsymbol{I}_{3\times3}(\pi/180)^2\boldsymbol{I}_{3\times3}[2\times(\pi/(180\times3600))]^2\boldsymbol{I}_{3\times3}\ [2\times(\pi/(180\times3600))]^2\boldsymbol{I}_{3\times3}5\boldsymbol{I}_{3\times3}0.02\boldsymbol{I}_{3\times3}1000\ 0.01\ 1\times10^{-4}\ 1\times10^{-4}]$$

仿真结果如图 4.1～图 4.8 所示。由图 4.1～图 4.2 可以看出，从星和主星的陀螺漂移均能得到很好的估计。图 4.3 和图 4.4 分别给出了相对四元数 $\boldsymbol{q}_{d/H}$ 和 $\boldsymbol{q}_{c/H}$ 的姿态估计误差与 3σ 界。图 4.5 和图 4.6 分别给出了相对位置和相对

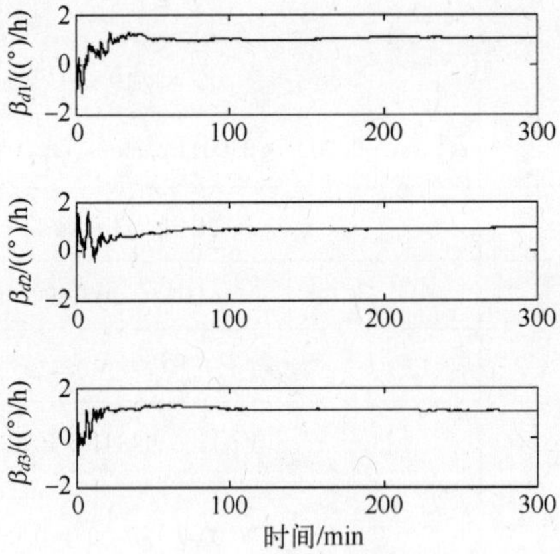

图 4.1　从星陀螺漂移估计

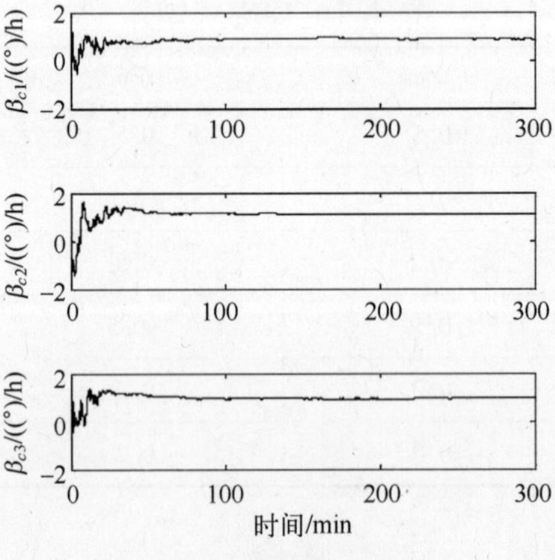

图 4.2　主星陀螺漂移估计

速度估计误差，估计误差始终控制在 3σ 界内。最终，三个方向的相对位置和速度误差均分别控制在 0.01m 和 2×10^{-5}m/s 以内。图 4.7 给出了两卫星间的相对姿态估计误差，单轴姿态估计精度在 0.02°以内，相对姿态可由式（2.29）计算得到。图 4.8 给出了主星轨道元素估计误差与 3σ 界。

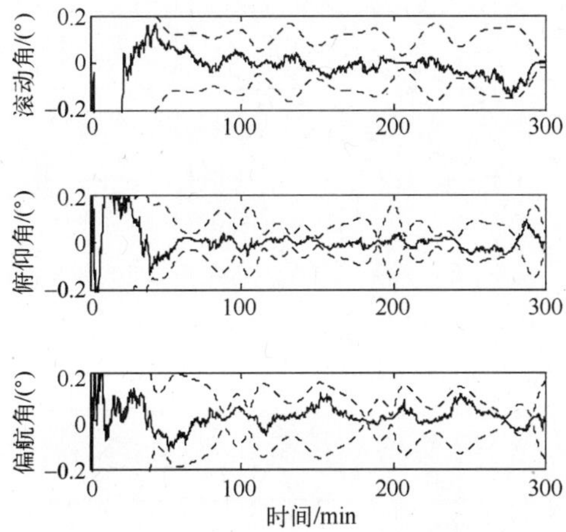

图 4.3　四元数 $\boldsymbol{q}_{d/H}$ 姿态估计误差与 3σ 界

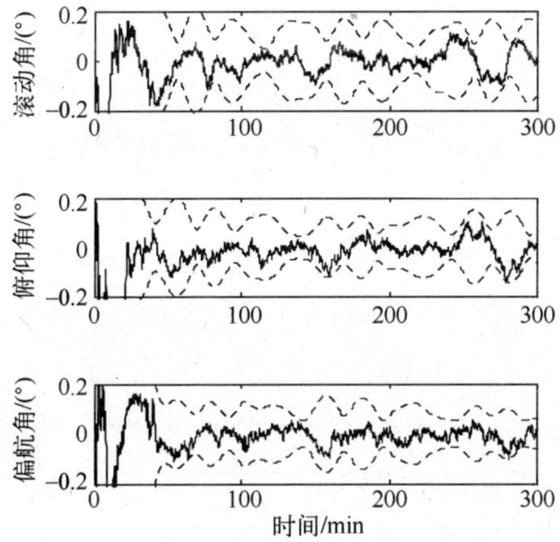

图 4.4　四元数 $\boldsymbol{q}_{c/H}$ 姿态估计误差与 3σ 界

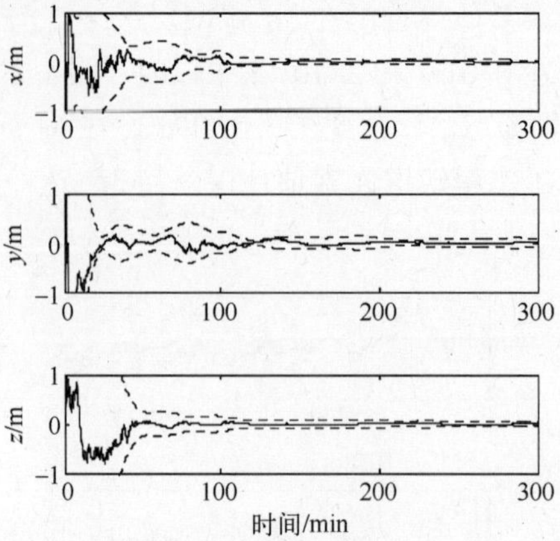

图 4.5　相对位置估计误差与 3σ 界

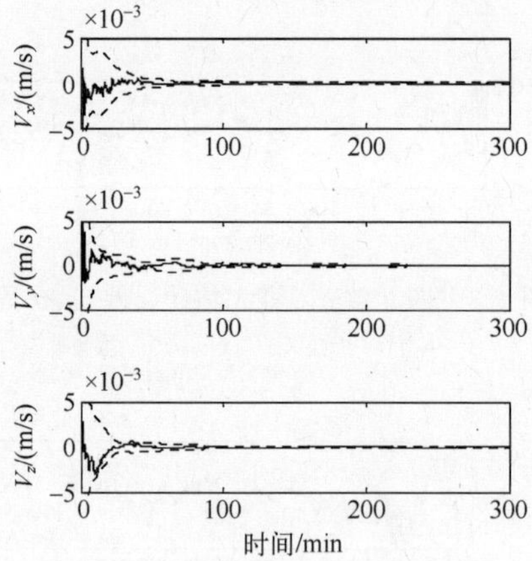

图 4.6　相对速度估计误差与 3σ 界

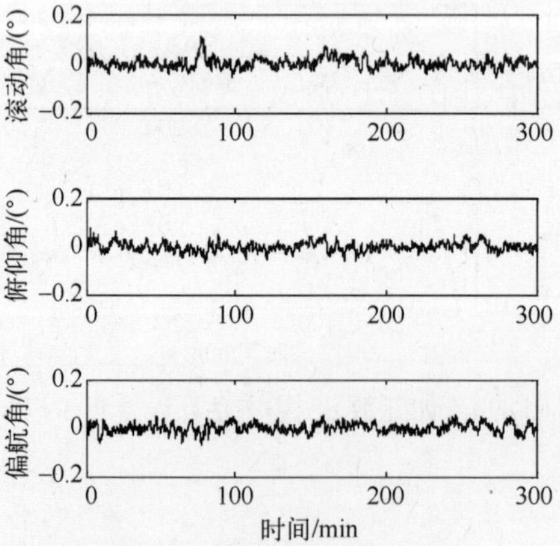

图 4.7　相对姿态估计误差

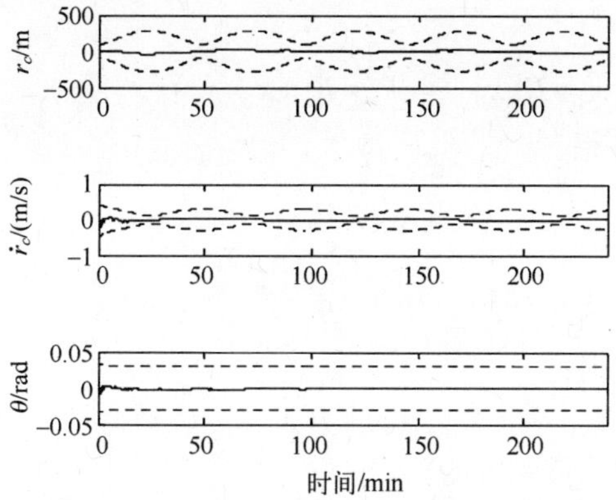

图 4.8　主星轨道元素估计误差与 3σ 界

4.2　基于 CKF 的航天器相对位姿确定方法

4.2.1　容积卡尔曼滤波

考虑具有加性噪声的非线性系统模型[8]：

$$\boldsymbol{x}_k = \boldsymbol{f}(\boldsymbol{x}_{k-1}, k) + \boldsymbol{G}_{k-1}\boldsymbol{w}_{k-1} \tag{4.48}$$

$$\tilde{\boldsymbol{z}}_k = \boldsymbol{h}(\boldsymbol{x}_k, k) + \boldsymbol{v}_k \tag{4.49}$$

式中：$\boldsymbol{x}_k \in \mathbb{R}^n$ 和 $\tilde{z}_k \in \mathbb{R}^m$ 分别为 k 时刻的状态矢量和观测矢量；$\boldsymbol{f}(\cdot)$ 和 $\boldsymbol{h}(\cdot)$ 为已知的非线性函数；$\boldsymbol{G}_{k-1}$ 为过程噪声分布矩阵；$\boldsymbol{w}_{k-1}$ 和 $\boldsymbol{v}_k$ 分别为动态系统的过程噪声和观测噪声，它们均为零均值高斯白噪声，且协方差阵分别为 $\boldsymbol{Q}_{k-1}$ 和 $\boldsymbol{R}_k$。

基于三阶容积法则，CKF 利用 $2n$ 个容积点加权求和来近似计算函数 $\boldsymbol{f}(\boldsymbol{x})$ 的多维积分：

$$\int_{\mathbb{R}^n} \boldsymbol{f}(\boldsymbol{x}) \mathcal{N}(\boldsymbol{x}; \boldsymbol{\mu}, \Sigma) \mathrm{d}\boldsymbol{x} \approx \frac{1}{2n} \sum_{i=1}^{2n} \boldsymbol{f}(\boldsymbol{\mu} + \sqrt{\Sigma}\xi_i) \tag{4.50}$$

式中：$\sqrt{\Sigma}$ 为协方差阵 Σ 的平方根，满足 $\Sigma = \sqrt{\Sigma}\sqrt{\Sigma}^{\mathrm{T}}$。容积点集 ξ_i 具有形式为

$$\xi_i = \begin{cases} \sqrt{n}\boldsymbol{e}_i, & i = 1, 2, \cdots, n \\ -\sqrt{n}\boldsymbol{e}_{i-n}, & i = n+1, n+2, \cdots, 2n \end{cases} \tag{4.51}$$

式中：$\boldsymbol{e}_i \in \mathbb{R}^n$ 为 $n \times n$ 单位阵的第 i 列。

整个算法流程如下：

1. 初始化

$$\hat{\boldsymbol{x}}_0=\boldsymbol{E}[\boldsymbol{x}_0],\quad \boldsymbol{P}_0=\boldsymbol{E}[(\boldsymbol{x}_0-\hat{\boldsymbol{x}}_0)(\boldsymbol{x}_0-\hat{\boldsymbol{x}}_0)^{\mathrm{T}}] \tag{4.52}$$

2. 时间更新方程

（1）计算容积点（$i=1,2,\cdots,2n$）：

$$\boldsymbol{P}_{k-1|k-1}=\boldsymbol{S}_{k-1|k-1}\boldsymbol{S}_{k-1|k-1}^{\mathrm{T}} \tag{4.53}$$

$$\boldsymbol{X}_{k-1|k-1}(i)=\boldsymbol{S}_{k-1|k-1}\xi_i+\hat{\boldsymbol{x}}_{k-1|k-1} \tag{4.54}$$

（2）容积点传播：

$$\boldsymbol{X}_{k|k-1}^{*}(i)=\boldsymbol{f}(\boldsymbol{X}_{k-1|k-1}(i)) \tag{4.55}$$

（3）计算预测均值和方差阵：

$$\hat{\boldsymbol{x}}_{k|k-1}=\frac{1}{2n}\sum_{i=1}^{2n}\boldsymbol{X}_{k|k-1}^{*}(i) \tag{4.56}$$

$$\boldsymbol{P}_{k|k-1}=\frac{1}{2n}\sum_{i=1}^{2n}\boldsymbol{X}_{k|k-1}^{*}(i)\boldsymbol{X}_{k|k-1}^{*\mathrm{T}}(i)-\hat{\boldsymbol{x}}_{k|k-1}\hat{\boldsymbol{x}}_{k|k-1}^{\mathrm{T}}+\boldsymbol{Q}_{k-1} \tag{4.57}$$

3. 观测更新方程

（1）计算容积点（$i=1,2,\cdots,2n$）：

$$\boldsymbol{P}_{k|k-1}=\boldsymbol{S}_{k|k-1}\boldsymbol{S}_{k|k-1}^{\mathrm{T}} \tag{4.58}$$

$$\boldsymbol{X}_{k|k-1}(i)=\boldsymbol{S}_{k|k-1}\xi_i+\hat{\boldsymbol{x}}_{k|k-1} \tag{4.59}$$

（2）容积点传播：

$$\boldsymbol{Z}_{k|k-1}(i)=\boldsymbol{h}(\boldsymbol{X}_{k|k-1}(i)) \tag{4.60}$$

（3）计算观测预测值、新息方差和协方差矩阵：

$$\hat{\boldsymbol{z}}_{k|k-1}=\frac{1}{2n}\sum_{i=1}^{2n}\boldsymbol{Z}_{k|k-1}(i) \tag{4.61}$$

$$\boldsymbol{P}_{zz,k|k-1}=\frac{1}{2n}\sum_{i=1}^{2n}\boldsymbol{Z}_{k|k-1}(i)\boldsymbol{Z}_{k|k-1}^{\mathrm{T}}(i)-\hat{\boldsymbol{z}}_{k|k-1}\hat{\boldsymbol{z}}_{k|k-1}^{\mathrm{T}}+\boldsymbol{R}_k \tag{4.62}$$

$$\boldsymbol{P}_{xz,k|k-1}=\frac{1}{2n}\sum_{i=1}^{2n}\boldsymbol{X}_{k|k-1}(i)\boldsymbol{Z}_{k|k-1}^{\mathrm{T}}(i)-\hat{\boldsymbol{x}}_{k|k-1}\hat{\boldsymbol{z}}_{k|k-1}^{\mathrm{T}} \tag{4.63}$$

4. 观测更新

$$\boldsymbol{K}_k=\boldsymbol{P}_{xz,k|k-1}\boldsymbol{P}_{zz,k|k-1}^{-1} \tag{4.64}$$

$$\hat{\boldsymbol{x}}_{k|k}=\hat{\boldsymbol{x}}_{k|k-1}+\boldsymbol{K}_k(\tilde{\boldsymbol{z}}_k-\hat{\boldsymbol{z}}_{k|k-1}) \tag{4.65}$$

$$\boldsymbol{P}_{k|k}=\boldsymbol{P}_{k|k-1}-\boldsymbol{K}_k\boldsymbol{P}_{zz,k|k-1}\boldsymbol{K}_k^{\mathrm{T}} \tag{4.66}$$

4.2.2 CKF 导航滤波器设计

4.2.2.1 CKF1

在所设计的 EKF 导航滤波器基础上，本节进行 CKF 导航滤波器设计。正如式（4.56）所示，计算预测四元数均值需要对一系列四元数容积点集进行加权平均，而所求得均值却无法保证满足四元数单位范数约束条件。因此，广义罗德里格参数（Generalized Rodrigues Parameters，GRPs）常被用来传播和更新四元数[9-10]。

广义罗德里格参数定义为[10]

$$\delta \boldsymbol{p} \equiv f\frac{\delta \boldsymbol{\varrho}}{a+\delta q_4} \tag{4.67}$$

式中：参数 a 从 0 到 1 取值；f 为标量因子。当 $a=0$，$f=1$ 时，式（4.67）退化为罗德里格参数（Rodrigues Parameters）或吉布斯矢量（Gibbs Vector）；当 $a=1$，$f=1$ 时，式（4.67）则退化为修正罗德里格参数（MRPs）。这里，选取 $f=2(a+1)$，以便在小角度误差情形下 $\|\delta \boldsymbol{p}\|$ 与旋转误差角大小近似相等。

定义状态变量及其容积点为

$$\hat{\boldsymbol{x}}_k^+ \equiv \begin{bmatrix} \delta\hat{\boldsymbol{p}}_{d/H_k}^+ \\ \delta\hat{\boldsymbol{p}}_{c/H_k}^+ \\ \hat{\boldsymbol{\beta}}_{d_k}^+ \\ \hat{\boldsymbol{\beta}}_{c_k}^+ \\ \hat{\boldsymbol{x}}_{p_k}^+ \end{bmatrix},\quad \boldsymbol{X}_k(i) \equiv \begin{bmatrix} \boldsymbol{X}_k^{\delta \boldsymbol{p}_{d/H}}(i) \\ \boldsymbol{X}_k^{\delta \boldsymbol{p}_{c/H}}(i) \\ \boldsymbol{X}_k^{\boldsymbol{\beta}_d}(i) \\ \boldsymbol{X}_k^{\boldsymbol{\beta}_c}(i) \\ \boldsymbol{X}_k^{\boldsymbol{x}_p}(i) \end{bmatrix},\quad \boldsymbol{X}_k^{\boldsymbol{x}_p}(i) \equiv \begin{bmatrix} \boldsymbol{X}_k^{\boldsymbol{\rho}}(i) \\ \boldsymbol{X}_k^{\dot{\boldsymbol{\rho}}}(i) \\ \boldsymbol{X}_k^{r_c}(i) \\ \boldsymbol{X}_k^{\dot{r}_c}(i) \\ \boldsymbol{X}_k^{\theta}(i) \\ \boldsymbol{X}_k^{\dot{\theta}}(i) \end{bmatrix},\quad i=1,2,\cdots,44 \tag{4.68}$$

式中：$\boldsymbol{X}_k^{\delta \boldsymbol{p}_{d/H}}$ 和 $\boldsymbol{X}_k^{\delta \boldsymbol{p}_{c/H}}$ 分别为相对四元数 $\boldsymbol{q}_{d/H}$ 和 $\boldsymbol{q}_{c/H}$ 的姿态误差部分；$\boldsymbol{X}_k^{\boldsymbol{\beta}_d}$ 和 $\boldsymbol{X}_k^{\boldsymbol{\beta}_c}$ 分别为从星和主星陀螺漂移部分；$\boldsymbol{X}_k^{\boldsymbol{x}_p}$ 为 10 维状态矢量 $\boldsymbol{x}_p$ 部分。为了传播 GRPs 容积点 $\boldsymbol{X}_k^{\delta \boldsymbol{p}_{d/H}}$ 和 $\boldsymbol{X}_k^{\delta \boldsymbol{p}_{c/H}}$，需要先定义四元数容积点。以 $\boldsymbol{X}_k^{\delta \boldsymbol{p}_{d/H}}$ 为例，首先计算四元数：

$$\hat{\boldsymbol{q}}_{d/H_k}^+(i) = \delta\hat{\boldsymbol{q}}_{d/H_k}^+(i) \otimes \hat{\boldsymbol{q}}_{d/H_k}^+,\quad i=1,2,\cdots,44 \tag{4.69}$$

式中：$\delta\hat{\boldsymbol{q}}_{d/H_k}^+(i) \equiv [\delta \boldsymbol{\varrho}_{d/H_k}^{+\mathrm{T}}(i)\ \ \delta q_{4,d/H_k}^+(i)]^{\mathrm{T}}$ 可以表示成

$$\delta q_{4,d/H_k}^+(i) = \frac{-a\|\boldsymbol{X}_k^{\delta \boldsymbol{p}_{d/H}}(i)\|^2 + f\sqrt{f^2+(1-a^2)\|\boldsymbol{X}_k^{\delta \boldsymbol{p}_{d/H}}(i)\|^2}}{f^2+\|\boldsymbol{X}_k^{\delta \boldsymbol{p}_{d/H}}(i)\|^2},\quad i=1,2,\cdots,44 \tag{4.70}$$

$$\delta \boldsymbol{\varrho}_{d/H_k}^+(i) = f^{-1}[a+\delta q_{4,d/H_k}^+(i)]\boldsymbol{X}_k^{\delta \boldsymbol{p}_{d/H}}(i),\quad i=1,2,\cdots,44 \tag{4.71}$$

接下来，式（4.69）中的四元数容积点可以用式（2.24）进行传播，即

$$\hat{\boldsymbol{q}}_{d/H_{k+1}}^-(i) = \overline{\Omega}(\hat{\boldsymbol{\omega}}_{d/I_k}^{d+}(i))\overline{\Gamma}(\hat{\boldsymbol{\omega}}_{H/I_k}^{H+}(i))\hat{\boldsymbol{q}}_{d/H_k}^+(i),\quad i=1,2,\cdots,44 \tag{4.72}$$

其中，对应的从星估计角速度为

$$\hat{\boldsymbol{\omega}}_{d/I_k}^{d+}(i)=\widetilde{\boldsymbol{\omega}}_{d/I_k}^{d}-\boldsymbol{X}_k^{\boldsymbol{\beta}_d}(i),\quad i=1,2,\cdots,44 \tag{4.73}$$

且主星 H 系的估计角速度为

$$\hat{\boldsymbol{\omega}}_{H/I_k}^{H+}(i)=[0\quad 0\quad \boldsymbol{X}_k^{\dot{\theta}}(i)]^{\mathrm{T}},\quad i=1,2,\cdots,44 \tag{4.74}$$

进一步，可得传播的误差四元数容积点

$$\delta\hat{\boldsymbol{q}}_{d/H_{k+1}}^{-}(i)=\hat{\boldsymbol{q}}_{d/H_{k+1}}^{-}(i)\otimes[\hat{\boldsymbol{q}}_{d/H_{k+1}}^{-}]^{-1},\quad i=1,2,\cdots,44 \tag{4.75}$$

其中，

$$\hat{\boldsymbol{q}}_{d/H_{k+1}}^{-}=\overline{\Omega}(\widetilde{\boldsymbol{\omega}}_{d/I_k}^{d}-\hat{\boldsymbol{\beta}}_{d_k}^{+})\overline{\Gamma}(\hat{\boldsymbol{\omega}}_{H/I_k}^{H+})\hat{\boldsymbol{q}}_{d/H_k}^{+} \tag{4.76}$$

最后，由式（4.67）计算得到传播的 GRPs 容积点为

$$\boldsymbol{X}_{k+1}^{\delta\boldsymbol{p}_{d/H}}(i)=f\frac{\delta\boldsymbol{\varrho}_{d/H_{k+1}}^{-}(i)}{a+\delta q_{4,d/H_{k+1}}^{-}(i)},\quad i=1,2,\cdots,44 \tag{4.77}$$

式中：$[\delta\boldsymbol{\varrho}_{d/H_{k+1}}^{-\mathrm{T}}(i)\ \delta q_{4,d/H_{k+1}}^{-}(i)]^{\mathrm{T}}=\delta\hat{\boldsymbol{q}}_{d/H_{k+1}}^{-}(i)$。类似地，GRPs 容积点 $\boldsymbol{X}_k^{\delta\boldsymbol{p}_{c/H}}$ 可采用同样的方式进行传播。

由于陀螺漂移斜率噪声是零均值高斯白噪声，所以有

$$\boldsymbol{X}_{k+1}^{\boldsymbol{\beta}_d}(i)=\boldsymbol{X}_k^{\boldsymbol{\beta}_d}(i),\quad \boldsymbol{X}_{k+1}^{\boldsymbol{\beta}_c}(i)=\boldsymbol{X}_k^{\boldsymbol{\beta}_c}(i),\quad i=1,2,\cdots,44 \tag{4.78}$$

最后，剩下的来自 $\boldsymbol{x}_p$ 的容积点可通过传播式（2.5）得到。因此，可利用式（4.56）和式（4.57）计算预测状态和方差阵。传播的四元数和相对位置容积点被用来计算观测预测值容积点，即

$$\boldsymbol{Z}_{k+1}(i)=\begin{bmatrix}\boldsymbol{A}_H^d(\hat{\boldsymbol{q}}_{d/H_{k+1}}^{-}(i))\hat{\boldsymbol{r}}_{1_{k+1}}^{-}(i)\\ \boldsymbol{A}_H^d(\hat{\boldsymbol{q}}_{d/H_{k+1}}^{-}(i))\hat{\boldsymbol{r}}_{2_{k+1}}^{-}(i)\\ \vdots\\ \boldsymbol{A}_H^d(\hat{\boldsymbol{q}}_{d/H_{k+1}}^{-}(i))\hat{\boldsymbol{r}}_{N_{k+1}}^{-}(i)\end{bmatrix},\quad i=1,2,\cdots,44 \tag{4.79}$$

其中，

$$\hat{\boldsymbol{r}}_{j_{k+1}}^{-}(i)=\frac{\boldsymbol{\mathcal{X}}_{j_{k+1}}'(i)-\boldsymbol{X}_{k+1}^{\boldsymbol{\rho}}(i)}{\|\boldsymbol{\mathcal{X}}_{j_{k+1}}'(i)-\boldsymbol{X}_{k+1}^{\boldsymbol{\rho}}(i)\|},\quad j=1,2,\cdots,N,\quad i=1,2,\cdots,44 \tag{4.80}$$

$$\boldsymbol{\mathcal{X}}_{j_{k+1}}'(i)=[\boldsymbol{A}_H^c(\hat{\boldsymbol{q}}_{c/H_{k+1}}^{-}(i))]^{\mathrm{T}}\boldsymbol{\mathcal{X}}_j,\quad j=1,2,\cdots,N,\quad i=1,2,\cdots,44 \tag{4.81}$$

于是，可利用式（4.61）~式（4.63）计算观测预测值、新息方差以及协方差矩阵。进一步，状态变量和方差矩阵可由式（4.65）和式（4.66）进行更新，更新的状态变量为 $\hat{\boldsymbol{x}}_{k+1}^{+}\equiv[\delta\hat{\boldsymbol{p}}_{d/H_{k+1}}^{+\mathrm{T}}\ \delta\hat{\boldsymbol{p}}_{c/H_{k+1}}^{+\mathrm{T}}\ \hat{\boldsymbol{\beta}}_{d_{k+1}}^{+\mathrm{T}}\ \hat{\boldsymbol{\beta}}_{c_{k+1}}^{+\mathrm{T}}\ \hat{\boldsymbol{x}}_{p_{k+1}}^{+\mathrm{T}}]^{\mathrm{T}}$。然后，利用式（4.70）和式（4.71）将 $\delta\hat{\boldsymbol{p}}_{d/H_{k+1}}^{+}$ 和 $\delta\hat{\boldsymbol{p}}_{c/H_{k+1}}^{+}$ 分别转换为 $\delta\hat{\boldsymbol{q}}_{d/H_{k+1}}^{+}$ 和 $\delta\hat{\boldsymbol{q}}_{c/H_{k+1}}^{+}$，用以计算更新的四元数：

$$\hat{\boldsymbol{q}}^{+}_{d/H_{k+1}}=\delta\hat{\boldsymbol{q}}^{+}_{d/H_{k+1}}\otimes\hat{\boldsymbol{q}}^{-}_{d/H_{k+1}},\quad \hat{\boldsymbol{q}}^{+}_{c/H_{k+1}}=\delta\hat{\boldsymbol{q}}^{+}_{c/H_{k+1}}\otimes\hat{\boldsymbol{q}}^{-}_{c/H_{k+1}} \tag{4.82}$$

值得注意的是，在下一次传播前需将 $\delta\hat{\boldsymbol{p}}^{+}_{d/H_{k+1}}$ 和 $\delta\hat{\boldsymbol{p}}^{+}_{c/H_{k+1}}$ 重置为零。

4.2.2.2　CKF2

最近，Chang 等指出式（4.75）中参考四元数最好的选择不是四元数 $\hat{\boldsymbol{q}}^{-}_{d/H_{k+1}}$ 而应是传播的四元数容积点集的加权平均值[11]。事实上，如果可能，参考四元数的最好选择应该是当前四元数的真值。显然，这是不现实的，因此，越接近四元数真值的参考四元数越受青睐。这里，四元数 $\hat{\boldsymbol{q}}^{-}_{d/H_{k+1}}$ 可认为是状态均值的线性传播，而传播的四元数容积点集的加权平均值比其中任何一个四元数容积点都更为准确。因此，传播的四元数容积点集的均值应该用来替换式（4.75）和式（4.82）中的 $\hat{\boldsymbol{q}}^{-}_{d/H_{k+1}}$。据参考文献［11］报道，这能够在一定程度上提高滤波性能，尤其是当动力学模型强非线性和对状态变量缺乏较好的先验估计信息时。因此，问题的关键在于如何加权平均这一系列四元数容积点集并维持四元数的单位范数属性。基于参考文献［12］中的加权平均思想，Markley 等进一步提出了包括标量加权和矩阵加权的最优四元数平均算法[13]。本节正是采用这一方法来改进先前提出的 CKF 导航滤波器的。

参考文献［13］指出，我们真正关心的是载体姿态而不是描述姿态的四元数。因此，可以通过对姿态阵进行加权平均从而求取最优加权平均四元数。构造代价函数：

$$\hat{\boldsymbol{q}}^{-}_{d/H_{k+1}} \triangleq \underset{\boldsymbol{q}\in\mathbb{S}^3}{\operatorname{argmin}}\sum_{i=1}^{2n} W_i^{\text{mean}}\ \|\boldsymbol{A}(\boldsymbol{q})-\boldsymbol{A}(\hat{\boldsymbol{q}}^{-}_{d/H_{k+1}}(i))\|_F^2 \tag{4.83}$$

式中：$\mathbb{S}^3$ 表示单位球；$\|\cdot\|_F$ 表示 Frobenius 范数；$W_i^{\text{mean}}=\dfrac{1}{2n}$ 为第 i 个四元数容积点的权重；$\boldsymbol{A}(\boldsymbol{q})$ 为对应于待求平均四元数 $\boldsymbol{q}$ 的姿态阵。可以看出，式（4.83）的意义在于使所计算得到的平均姿态阵与样本姿态阵之间距离的均方和在 Frobenius 范数空间内最小。进一步，平均四元数 $\hat{\boldsymbol{q}}^{-}_{d/H_{k+1}}$ 的求取可以转换为求解最大值问题[13]，即

$$\hat{\boldsymbol{q}}^{-}_{d/H_{k+1}} \triangleq \underset{\boldsymbol{q}\in\mathbb{S}^3}{\operatorname{argmax}}\ \boldsymbol{q}^{\mathrm{T}}\boldsymbol{M}\boldsymbol{q} \tag{4.84}$$

$$\boldsymbol{M}=\sum_{i=1}^{2n} W_i^{\text{mean}}\ \hat{\boldsymbol{q}}^{-}_{d/H_{k+1}}(i)\ \hat{\boldsymbol{q}}^{-}_{d/H_{k+1}}(i)^{\mathrm{T}} \tag{4.85}$$

平均四元数 $\hat{\boldsymbol{q}}^{-}_{d/H_{k+1}}$ 就是矩阵 $\boldsymbol{M}$ 的最大特征值所对应的特征矢量，所有求解 Wahba 问题的方法均可用来计算该加权平均四元数。参考文献［13］进一步指出该解为最小均方误差（Minimum Mean Square Error，MMSE）意义上的最优解，因此所得加权平均四元数为最优平均四元数。另外，利用式（4.84）

求解加权平均四元数还具有两个优点：①改变任一$\hat{\boldsymbol{q}}^{-}_{d/H_{k+1}}(i)$的符号不改变均值；②解出的平均四元数满足正交性约束条件。

同样地，可计算得到最优加权平均四元数$\hat{\boldsymbol{q}}^{-}_{c/H_{k+1}}$。最后，将所求得最优平均四元数替换式（4.75）和式（4.82）中的$\hat{\boldsymbol{q}}^{-}_{d/H_{k+1}}$和$\hat{\boldsymbol{q}}^{-}_{c/H_{k+1}}$，这就是先前提出的 CKF 导航滤波器的改进版本。

4.2.3 仿真实验与分析

本节对比了 CKF 与 EKF 导航滤波器在具有初始条件误差情形下的滤波性能。仿真参数设置与4.1.3 节基本一致。在4.1.3 节中，除了陀螺漂移，其余状态量的滤波初值均为真值。在本节数值仿真过程中，将会对这些变量引入初始条件误差。对于相对姿态四元数$\boldsymbol{q}_{d/H}$和$\boldsymbol{q}_{c/H}$，分别加上初始姿态误差$[10° \ -10° \ 5°]^{\mathrm{T}}$和$[-10° \ 10° \ 5°]^{\mathrm{T}}$；从星和主星的陀螺漂移初值仍设为零；相对位置和速度、主星的轨道半径、径向速率、真近点角以及真近点角速率滤波初值均在真值上加上 1σ 误差。这些状态量的初始方差与4.1.3 节一致。对于初始姿态误差，其初始方差可设为$[10\times(\pi/180)]^2\ \mathrm{rad}^2$。因此，滤波初始方差阵为

$$\boldsymbol{P}_0=\mathrm{blkdiag}[(10\times\pi/180)^2\boldsymbol{I}_{3\times3}(10\times\pi/180)^2\boldsymbol{I}_{3\times3}[2\times(\pi/(180\times3600))]^2\boldsymbol{I}_{3\times3}$$
$$[2\times(\pi/(180\times3600))]^2\boldsymbol{I}_{3\times3}\ 5\boldsymbol{I}_{3\times3}\ 0.02\boldsymbol{I}_{3\times3}\ 1000\ 0.01\ 1\times10^{-4}\ 1\times10^{-4}]$$

经过 50 次蒙特卡洛仿真，图 4.9～图 4.16 对比了 EKF 和两种 CKF 导航滤波器性能。图4.9 和图4.10 分别给出了相对四元数$\boldsymbol{q}_{d/H}$和$\boldsymbol{q}_{c/H}$的姿态估计误差。两种 CKF 估计误差均能在 100min 内收敛到各自的 3σ 界内，而 EKF 估计误差无法收敛。图 4.11 和图 4.12 分别给出了相对位置和速度的估计误差范数。可以看出，EKF 估计效果非常差且发散，这表明一阶近似无法捕获初始条件误差；而 CKF 估计效果较好，最终相对位置和速度估计误差范数分别为 0.03m 和 3×10^{-5}m/s。局部放大图表明 CKF2 比 CKF1 滤波性能略优，这与前面的分析一致。图 4.13 和图 4.14 分别给出了从星和主星陀螺漂移估计误差范数。图 4.15 给出了两卫星间的相对姿态误差范数。可以看出，CKF1 和 CKF2 的姿态估计精度均能在 300min 内收敛到 0.05°以内，而 EKF 无法收敛。图 4.16 给出了 CKF1 和 CKF2 的主星轨道元素估计误差，而由于发散严重就没有对 EKF 估计误差进行绘制。

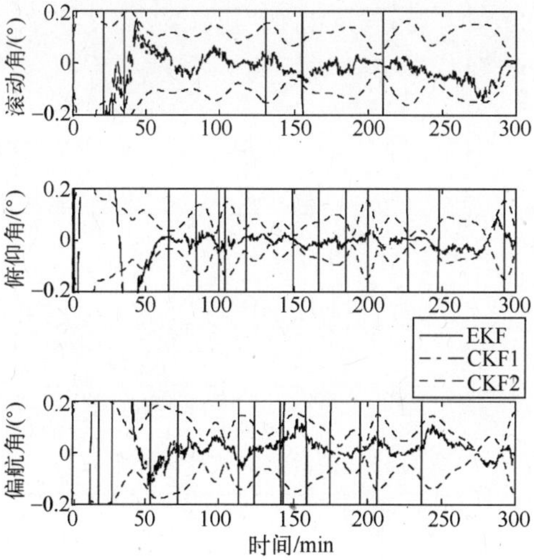

图 4.9　四元数 $\boldsymbol{q}_{d/H}$ 姿态估计误差与 3σ 界

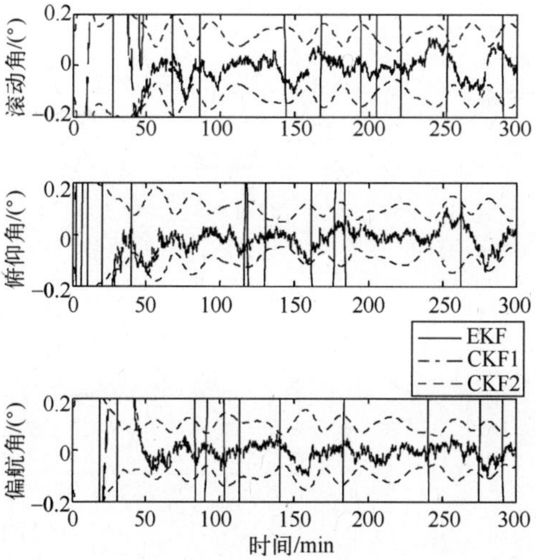

图 4.10　四元数 $\boldsymbol{q}_{c/H}$ 姿态估计误差与 3σ 界

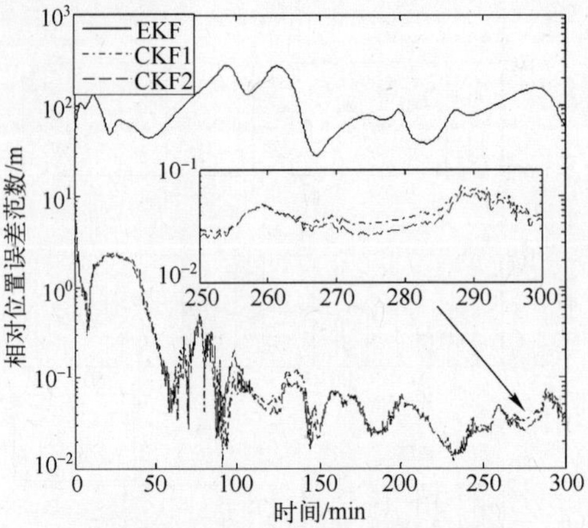

图 4.11　相对位置估计误差范数

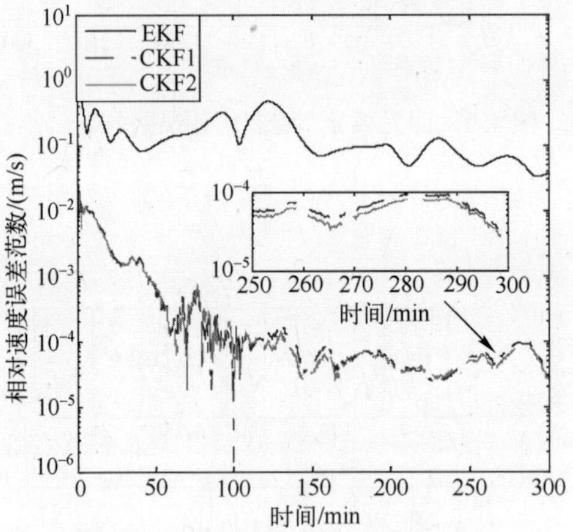

图 4.12　相对速度估计误差范数

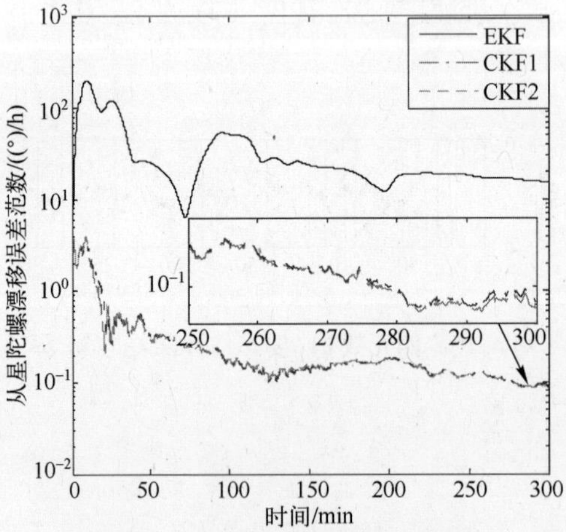

图 4.13　从星陀螺漂移误差范数

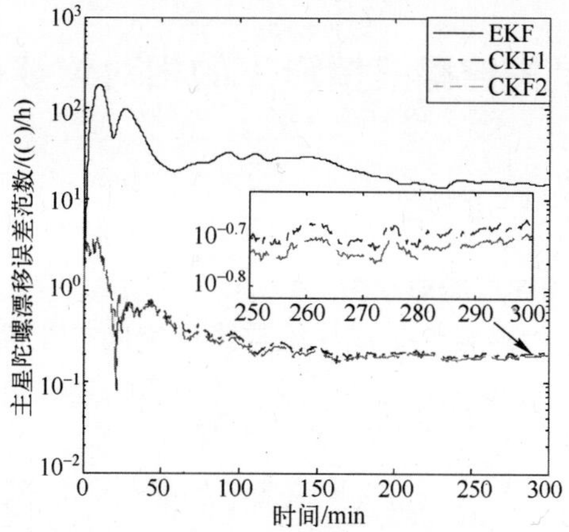

图 4.14　主星陀螺漂移误差范数

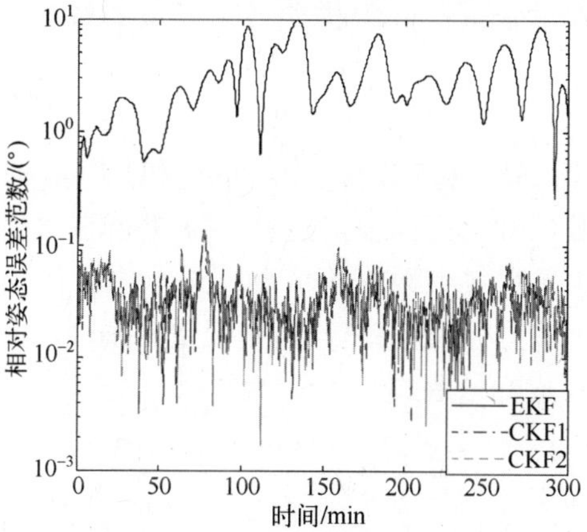

图 4.15　相对姿态估计误差范数

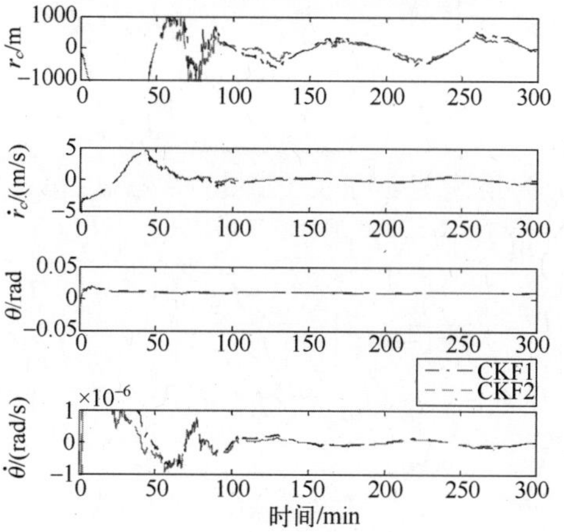

图 4.16　主星轨道元素估计误差

4.3 主星失控翻滚情形下的相对位姿确定方法

4.3.1 角速度测量模型

4.3.1.1 主星角速度测量模型

本节假设主星由于能料耗尽或系统故障在空间处于翻滚状态，且与从星之间无法进行通信联系。对于主星而言，其姿态角速度需要通过姿态动力学方程传播得到。

根据欧拉动力学方程，主星失控情形下的姿态动力学方程可以描述为

$$\dot{\boldsymbol{\omega}}_{c/I}^{c}=-\boldsymbol{J}_{c}^{-1}(\boldsymbol{\omega}_{c/I}^{c}\times\boldsymbol{J}_{c}\boldsymbol{\omega}_{c/I}^{c})+\boldsymbol{J}_{c}^{-1}\boldsymbol{d}_{c} \tag{4.86}$$

式中：$\boldsymbol{J}_c$ 为主星的惯量张量；$\boldsymbol{d}_c$ 为扰动力矩矢量。将式（4.86）对 $\boldsymbol{\omega}_{c/I}^{c}$ 线性化可得

$$\Delta\dot{\boldsymbol{\omega}}_{c/I}^{c}=\boldsymbol{J}_{c}^{-1}([\boldsymbol{J}_{c}\boldsymbol{\omega}_{c/I}^{c}\times]-[\boldsymbol{\omega}_{c/I}^{c}\times]\boldsymbol{J}_{c})\Delta\boldsymbol{\omega}_{c/I}^{c}+\boldsymbol{J}_{c}^{-1}\boldsymbol{d}_{c} \tag{4.87}$$

若惯量参数精确已知，则利用式（4.86）和式（4.87）可直接传播主星角速度；若惯量参数未知，则需要对这些参数进行在线估计。这里，考虑后一种情形，即缺少主星惯量参数先验信息。选取主星的惯量主轴坐标系为本体坐标系（同2.1.1节坐标系定义），且主惯量分别表示为 J_{cxx}，J_{cyy} 和 J_{czz}。那么，式（4.86）可写成标量形式

$$\begin{cases}\dot{\omega}_{cx}=\dfrac{(J_{cyy}-J_{czz})\omega_{cy}\omega_{cz}}{J_{cxx}}+\dfrac{d_{cx}}{J_{cxx}}\\ \dot{\omega}_{cy}=\dfrac{(J_{czz}-J_{cxx})\omega_{cx}\omega_{cz}}{J_{cyy}}+\dfrac{J_{cxx}}{J_{cyy}}\dfrac{d_{cy}}{J_{cxx}}\\ \dot{\omega}_{cz}=\dfrac{(J_{cxx}-J_{cyy})\omega_{cx}\omega_{cy}}{J_{czz}}+\dfrac{J_{cxx}}{J_{czz}}\dfrac{d_{cz}}{J_{cxx}}\end{cases} \tag{4.88}$$

式中：ω_{cx}，ω_{cy} 和 ω_{cz} 为角速度 $\boldsymbol{\omega}_{c/I}^{c}$ 的三分量；d_{cx}，d_{cy} 和 d_{cz} 为扰动力矩 $\boldsymbol{d}_c$ 的三分量。定义惯量比矢量 $\boldsymbol{l}=[l_x\ l_y\ l_z]^{\mathrm{T}}$ 为

$$l_x=\frac{J_{cyy}-J_{czz}}{J_{cxx}},\quad l_y=\frac{J_{czz}-J_{cxx}}{J_{cyy}},\quad l_z=\frac{J_{cxx}-J_{cyy}}{J_{czz}} \tag{4.89}$$

和新扰动力矩矢量为 $\boldsymbol{d}_c'=[d_{cx}/J_{cxx}\ d_{cy}/J_{cxx}\ d_{cz}/J_{cxx}]^{\mathrm{T}}$。进一步，由式（4.89）可得

$$\frac{J_{cxx}}{J_{cyy}}=\frac{1-l_y}{1+l_x},\quad \frac{J_{cxx}}{J_{czz}}=\frac{1+l_z}{1-l_x} \tag{4.90}$$

将式（4.89）和式（4.90）代入式（4.88），则主星的姿态动力学方程可重新描述成

$$\dot{\boldsymbol{\omega}}_{c/I}^{c}=\boldsymbol{\psi}(\boldsymbol{\omega}_{c/I}^{c},\boldsymbol{l})+\boldsymbol{J}(\boldsymbol{l})\boldsymbol{d}_{c}' \tag{4.91}$$

其中，

$$\boldsymbol{\psi}(\boldsymbol{\omega}_{c/I}^{c},\boldsymbol{l})=\begin{bmatrix} l_x\omega_{cy}\omega_{cz} \\ l_y\omega_{cx}\omega_{cz} \\ l_z\omega_{cx}\omega_{cy} \end{bmatrix},\quad \boldsymbol{J}(\boldsymbol{l})=\begin{bmatrix} 1 & 0 & 0 \\ 0 & \dfrac{1-l_y}{1+l_x} & 0 \\ 0 & 0 & \dfrac{1+l_z}{1-l_x} \end{bmatrix} \tag{4.92}$$

线性化方程（4.91）可得

$$\Delta\dot{\boldsymbol{\omega}}_{c/I}^{c}=\boldsymbol{M}(\boldsymbol{\omega}_{c/I}^{c},\boldsymbol{l})\Delta\boldsymbol{\omega}_{c/I}^{c}+\boldsymbol{N}(\boldsymbol{\omega}_{c/I}^{c})\Delta\boldsymbol{l}+\boldsymbol{J}(\boldsymbol{l})\boldsymbol{d}_{c}' \tag{4.93}$$

其中，

$$\boldsymbol{M}(\boldsymbol{\omega}_{c/I}^{c},\boldsymbol{l})=\frac{\partial\boldsymbol{\psi}}{\partial\boldsymbol{\omega}_{c/I}^{c}}=\begin{bmatrix} 0 & l_x\omega_{cz} & l_x\omega_{cy} \\ l_y\omega_{cz} & 0 & l_y\omega_{cx} \\ l_z\omega_{cy} & l_z\omega_{cx} & 0 \end{bmatrix} \tag{4.94}$$

$$\boldsymbol{N}(\boldsymbol{\omega}_{c/I}^{c})=\frac{\partial\boldsymbol{\psi}}{\partial\boldsymbol{l}}=\begin{bmatrix} \omega_{cy}\omega_{cz} & 0 & 0 \\ 0 & \omega_{cx}\omega_{cz} & 0 \\ 0 & 0 & \omega_{cx}\omega_{cy} \end{bmatrix} \tag{4.95}$$

4.3.1.2　从星角速度测量模型

从星上装有陀螺仪可直接进行角速度测量，相关数学模型详见 2.3.2 节。

4.3.2　EKF 导航滤波器设计

4.3.2.1　系统状态方程及其线性化

结合 4.1.2 节与前述的角速度测量模型，定义状态变量

$$\boldsymbol{x}\equiv[\boldsymbol{q}_{d/H}^{\mathrm{T}}\quad \boldsymbol{q}_{c/H}^{\mathrm{T}}\quad \boldsymbol{\beta}_{d}^{\mathrm{T}}\quad \boldsymbol{\omega}_{c/I}^{c\mathrm{T}}\quad \boldsymbol{l}^{\mathrm{T}}\quad \boldsymbol{\rho}^{\mathrm{T}}\quad \dot{\boldsymbol{\rho}}^{\mathrm{T}}\quad r_c\quad \dot{r}_c\quad \theta\quad \dot{\theta}]^{\mathrm{T}} \tag{4.96}$$

和误差状态变量

$$\Delta\boldsymbol{x}\equiv[\delta\boldsymbol{\alpha}_{d/H}^{\mathrm{T}}\quad \delta\boldsymbol{\alpha}_{c/H}^{\mathrm{T}}\quad \Delta\boldsymbol{\beta}_{d}^{\mathrm{T}}\quad \Delta\boldsymbol{\omega}_{c/I}^{c\mathrm{T}}\quad \Delta\boldsymbol{l}^{\mathrm{T}}\quad \Delta\boldsymbol{\rho}^{\mathrm{T}}\quad \Delta\dot{\boldsymbol{\rho}}^{\mathrm{T}}\quad \Delta r_c\quad \Delta\dot{r}_c\quad \Delta\theta\quad \Delta\dot{\theta}]^{\mathrm{T}} \tag{4.97}$$

此时，系统状态估计方程为

$$\dot{\hat{\boldsymbol{q}}}_{d/H}=\frac{1}{2}\Xi(\hat{\boldsymbol{q}}_{d/H})\hat{\boldsymbol{\omega}}_{d/H}^{d} \tag{4.98}$$

$$\hat{\boldsymbol{\omega}}_{d/H}^{d}=\hat{\boldsymbol{\omega}}_{d/I}^{d}-\boldsymbol{A}_{H}^{d}(\hat{\boldsymbol{q}}_{d/H})\hat{\boldsymbol{\omega}}_{H/I}^{H} \tag{4.99}$$

$$\dot{\hat{\boldsymbol{q}}}_{c/H}=\frac{1}{2}\Xi(\hat{\boldsymbol{q}}_{c/H})\hat{\boldsymbol{\omega}}_{c/H}^{c} \tag{4.100}$$

$$\hat{\boldsymbol{\omega}}_{c/H}^{c}=\hat{\boldsymbol{\omega}}_{c/I}^{c}-\boldsymbol{A}_{H}^{c}(\hat{\boldsymbol{q}}_{c/H})\hat{\boldsymbol{\omega}}_{H/I}^{H} \tag{4.101}$$

$$\hat{\boldsymbol{\omega}}_{d/I}^{d}=\widetilde{\boldsymbol{\omega}}_{d/I}^{d}-\hat{\boldsymbol{\beta}}_{d} \tag{4.102}$$

$$\dot{\hat{\boldsymbol{\omega}}}_{c/I}^{c}=\boldsymbol{\psi}(\hat{\boldsymbol{\omega}}_{c/I}^{c},\hat{\boldsymbol{l}}) \tag{4.103}$$

$$\dot{\hat{\boldsymbol{x}}}_{p}=\boldsymbol{f}(\hat{\boldsymbol{x}}_{p}) \tag{4.104}$$

$$\hat{\boldsymbol{\omega}}_{H/I}^{H}=[0\ \ 0\ \ \dot{\hat{\theta}}\ ]^{\mathrm{T}} \tag{4.105}$$

$$\dot{\hat{\boldsymbol{\beta}}}_{d}=\boldsymbol{0} \tag{4.106}$$

$$\dot{\hat{\boldsymbol{l}}}=\boldsymbol{0} \tag{4.107}$$

采用类似于4.1.2节中的线性化方法，可得如下误差状态方程：

$$\Delta\dot{\boldsymbol{x}}=\boldsymbol{F}\Delta\boldsymbol{x}+\boldsymbol{G}\boldsymbol{w} \tag{4.108}$$

其中，

$$\boldsymbol{F}\equiv\begin{bmatrix}-[\hat{\boldsymbol{\omega}}_{d/I}^{d}\times] & \boldsymbol{0}_{3\times3} & -\boldsymbol{I}_{3\times3} & \boldsymbol{0}_{3\times3} & \boldsymbol{0}_{3\times3} & \boldsymbol{0}_{3\times9}\ -\boldsymbol{A}_{H}^{d}(\hat{\boldsymbol{q}}_{d/H})\boldsymbol{n}\\ \boldsymbol{0}_{3\times3} & -[\hat{\boldsymbol{\omega}}_{c/I}^{c}\times] & \boldsymbol{0}_{3\times3} & \boldsymbol{I}_{3\times3} & \boldsymbol{0}_{3\times3} & \boldsymbol{0}_{3\times9}\ -\boldsymbol{A}_{H}^{c}(\hat{\boldsymbol{q}}_{c/H})\boldsymbol{n}\\ \boldsymbol{0}_{3\times3} & \boldsymbol{0}_{3\times3} & \boldsymbol{0}_{3\times3} & \boldsymbol{0}_{3\times3} & \boldsymbol{0}_{3\times3} & \boldsymbol{0}_{3\times10}\\ \boldsymbol{0}_{3\times3} & \boldsymbol{0}_{3\times3} & \boldsymbol{0}_{3\times3} & \boldsymbol{M}(\hat{\boldsymbol{\omega}}_{c/I}^{c},\hat{\boldsymbol{l}}) & \boldsymbol{N}(\hat{\boldsymbol{\omega}}_{c/I}^{c}) & \boldsymbol{0}_{3\times10}\\ \boldsymbol{0}_{3\times3} & \boldsymbol{0}_{3\times3} & \boldsymbol{0}_{3\times3} & \boldsymbol{0}_{3\times3} & \boldsymbol{0}_{3\times3} & \boldsymbol{0}_{3\times10}\\ \boldsymbol{0}_{10\times3} & \boldsymbol{0}_{10\times3} & \boldsymbol{0}_{10\times3} & \boldsymbol{0}_{10\times3} & \boldsymbol{0}_{10\times3} & \left.\dfrac{\partial f(\boldsymbol{x}_{p})}{\partial\boldsymbol{x}_{p}}\right|_{\hat{\boldsymbol{x}}_{p}}\end{bmatrix},$$

$$\boldsymbol{n}=[0\ \ 0\ \ 1]^{\mathrm{T}} \tag{4.109}$$

$$\boldsymbol{G}\equiv\begin{bmatrix}-\boldsymbol{I}_{3\times3} & \boldsymbol{0}_{3\times3} & \boldsymbol{0}_{3\times3} & \boldsymbol{0}_{3\times3}\\ \boldsymbol{0}_{3\times3} & \boldsymbol{0}_{3\times3} & \boldsymbol{0}_{3\times3} & \boldsymbol{0}_{3\times3}\\ \boldsymbol{0}_{3\times3} & \boldsymbol{0}_{3\times3} & \boldsymbol{I}_{3\times3} & \boldsymbol{0}_{3\times3}\\ \boldsymbol{0}_{3\times3} & \boldsymbol{0}_{3\times3} & \boldsymbol{J}(\hat{\boldsymbol{l}}) & \boldsymbol{0}_{3\times3}\\ \boldsymbol{0}_{3\times3} & \boldsymbol{0}_{3\times3} & \boldsymbol{0}_{3\times3} & \boldsymbol{0}_{3\times3}\\ \boldsymbol{0}_{3\times3} & \boldsymbol{0}_{3\times3} & \boldsymbol{0}_{3\times3} & \boldsymbol{0}_{3\times3}\\ \boldsymbol{0}_{3\times3} & \boldsymbol{0}_{3\times3} & \boldsymbol{0}_{3\times3} & \boldsymbol{I}_{3\times3}\\ \boldsymbol{0}_{2\times3} & \boldsymbol{0}_{2\times3} & \boldsymbol{0}_{2\times3} & \boldsymbol{0}_{2\times3}\\ \boldsymbol{0}_{2\times3} & \boldsymbol{0}_{2\times3} & \boldsymbol{0}_{2\times3} & \boldsymbol{0}_{2\times3}\end{bmatrix} \tag{4.110}$$

$$\boldsymbol{w} \equiv [\boldsymbol{\eta}_{dv}^{\mathrm{T}} \quad \boldsymbol{\eta}_{du}^{\mathrm{T}} \quad \boldsymbol{d}_c'^{\mathrm{T}} \quad \boldsymbol{\varpi}^{\mathrm{T}}]^{\mathrm{T}} \tag{4.111}$$

过程噪声 $\boldsymbol{w}$ 的谱密度矩阵为

$$\boldsymbol{Q} = \begin{bmatrix} \sigma_{dv}^2 \boldsymbol{I}_{3\times3} & \boldsymbol{0}_{3\times3} & \boldsymbol{0}_{3\times3} & \boldsymbol{0}_{3\times3} \\ \boldsymbol{0}_{3\times3} & \sigma_{du}^2 \boldsymbol{I}_{3\times3} & \boldsymbol{0}_{3\times3} & \boldsymbol{0}_{3\times3} \\ \boldsymbol{0}_{3\times3} & \boldsymbol{0}_{3\times3} & \sigma_{d_c'}^2 \boldsymbol{I}_{3\times3} & \boldsymbol{0}_{3\times3} \\ \boldsymbol{0}_{3\times3} & \boldsymbol{0}_{3\times3} & \boldsymbol{0}_{3\times3} & \sigma_{\varpi}^2 \boldsymbol{I}_{3\times3} \end{bmatrix} \tag{4.112}$$

4.3.2.2　状态和方差传播

状态变量$\hat{\boldsymbol{x}}$可通过积分方程（4.98）~方程（4.107）得到，方差传播可参见 4.1.2.2 节。

4.3.2.3　观测方程及其线性化

经过一系列数学推导，可得 VISNAV 测量灵敏度矩阵为

$$\boldsymbol{H}_k(\hat{\boldsymbol{x}}_k^-) = \begin{bmatrix} [\boldsymbol{A}_H^d(\hat{\boldsymbol{q}}_{d/H}^-)\hat{\boldsymbol{r}}_1^- \times] & \dfrac{\partial \hat{\boldsymbol{b}}_1^-}{\partial \delta\boldsymbol{\alpha}_{c/H}} & \boldsymbol{0}_{3\times3} & \boldsymbol{0}_{3\times3} & \boldsymbol{0}_{3\times3} & \dfrac{\partial \hat{\boldsymbol{b}}_1^-}{\partial \hat{\boldsymbol{\rho}}^-} & \boldsymbol{0}_{3\times7} \\ \vdots & \vdots & \vdots & \vdots & \vdots & \vdots & \vdots \\ [\boldsymbol{A}_H^d(\hat{\boldsymbol{q}}_{d/H}^-)\hat{\boldsymbol{r}}_N^- \times] & \dfrac{\partial \hat{\boldsymbol{b}}_N^-}{\partial \delta\boldsymbol{\alpha}_{c/H}} & \boldsymbol{0}_{3\times3} & \boldsymbol{0}_{3\times3} & \boldsymbol{0}_{3\times3} & \dfrac{\partial \hat{\boldsymbol{b}}_N^-}{\partial \hat{\boldsymbol{\rho}}^-} & \boldsymbol{0}_{3\times7} \end{bmatrix} \tag{4.113}$$

式中：$\hat{\boldsymbol{r}}_i^-$，$\partial \hat{\boldsymbol{b}}_i^- / \partial \delta\boldsymbol{\alpha}_{c/H}$和$\partial \hat{\boldsymbol{b}}_i^- / \partial \hat{\boldsymbol{\rho}}^-$已经在 4.1.2.3 节中进行过推导，这里不再赘述。表 4.5 概括了主星失控翻滚情形下的相对位姿确定算法。

表 4.5　主星失控翻滚情形下的相对位姿确定算法

初　始　化	$\hat{\boldsymbol{x}}(t_0) = \hat{\boldsymbol{x}}_0, \boldsymbol{P}(t_0) = \boldsymbol{P}_0$
增益及测量灵敏度矩阵	$\boldsymbol{K}_k = \boldsymbol{P}_k^- \boldsymbol{H}_k^{\mathrm{T}}(\hat{\boldsymbol{x}}_k^-)[\boldsymbol{H}_k(\hat{\boldsymbol{x}}_k^-)\boldsymbol{P}_k^- \boldsymbol{H}_k^{\mathrm{T}}(\hat{\boldsymbol{x}}_k^-) + \boldsymbol{R}_k]^{-1}$ $\boldsymbol{H}_k(\hat{\boldsymbol{x}}_k^-) = \left.\begin{bmatrix} [\boldsymbol{A}_H^d(\hat{\boldsymbol{q}}_{d/H}^-)\hat{\boldsymbol{r}}_1^- \times] & \dfrac{\partial \hat{\boldsymbol{b}}_1^-}{\partial \delta\boldsymbol{\alpha}_{c/H}} & \boldsymbol{0}_{3\times3} & \boldsymbol{0}_{3\times3} & \boldsymbol{0}_{3\times3} & \dfrac{\partial \hat{\boldsymbol{b}}_1^-}{\partial \hat{\boldsymbol{\rho}}^-} & \boldsymbol{0}_{3\times7} \\ \vdots & \vdots & \vdots & \vdots & \vdots & \vdots & \vdots \\ [\boldsymbol{A}_H^d(\hat{\boldsymbol{q}}_{d/H}^-)\hat{\boldsymbol{r}}_N^- \times] & \dfrac{\partial \hat{\boldsymbol{b}}_N^-}{\partial \delta\boldsymbol{\alpha}_{c/H}} & \boldsymbol{0}_{3\times3} & \boldsymbol{0}_{3\times3} & \boldsymbol{0}_{3\times3} & \dfrac{\partial \hat{\boldsymbol{b}}_N^-}{\partial \hat{\boldsymbol{\rho}}^-} & \boldsymbol{0}_{3\times7} \end{bmatrix}\right\|_{t_k}$ $\hat{\boldsymbol{r}}_i^- = \dfrac{[\boldsymbol{A}_H^c(\hat{\boldsymbol{q}}_{c/H}^-)]^{\mathrm{T}}\boldsymbol{\chi}_i - \hat{\boldsymbol{\rho}}^-}{\Vert [\boldsymbol{A}_H^c(\hat{\boldsymbol{q}}_{c/H}^-)]^{\mathrm{T}}\boldsymbol{\chi}_i - \hat{\boldsymbol{\rho}}^- \Vert}, \quad i = 1, 2, \cdots, N$

续表

初 始 化	$\hat{\boldsymbol{x}}(t_0)=\hat{\boldsymbol{x}}_0,\boldsymbol{P}(t_0)=\boldsymbol{P}_0$
更新	$\boldsymbol{P}_k^+=[\boldsymbol{I}-\boldsymbol{K}_k\boldsymbol{H}_k(\hat{\boldsymbol{x}}_k^-)]\boldsymbol{P}_k^-,\quad \Delta\hat{\boldsymbol{x}}_k^+=\boldsymbol{K}_k[\tilde{\boldsymbol{z}}_k-\boldsymbol{h}(\hat{\boldsymbol{x}}_k^-)]$ $\Delta\hat{\boldsymbol{x}}_k^+\equiv[\delta\hat{\boldsymbol{\alpha}}_{d/H_k}^{+\mathrm{T}}\ \delta\hat{\boldsymbol{\alpha}}_{c/H_k}^{+\mathrm{T}}\ \Delta\hat{\boldsymbol{\beta}}_{d_k}^{+\mathrm{T}}\ \Delta\hat{\boldsymbol{\omega}}_{c/I_k}^{c+\mathrm{T}}\ \Delta\hat{\boldsymbol{l}}_k^{+\mathrm{T}}\ \Delta\hat{\boldsymbol{x}}_{p_k}^{+\mathrm{T}}]^{\mathrm{T}}$ $\boldsymbol{h}_k(\hat{\boldsymbol{x}}_k^-)=\left.\begin{bmatrix}\boldsymbol{A}_H^d(\hat{\boldsymbol{q}}_{d/H}^-)\hat{\boldsymbol{r}}_1^-\\\boldsymbol{A}_H^d(\hat{\boldsymbol{q}}_{d/H}^-)\hat{\boldsymbol{r}}_2^-\\\vdots\\\boldsymbol{A}_H^d(\hat{\boldsymbol{q}}_{d/H}^-)\hat{\boldsymbol{r}}_N^-\end{bmatrix}\right\|_{t_k}$ $\hat{\boldsymbol{q}}_{d/H_k}^+=\hat{\boldsymbol{q}}_{d/H_k}^-+\frac{1}{2}\boldsymbol{\Xi}(\hat{\boldsymbol{q}}_{d/H_k}^-)\delta\hat{\boldsymbol{\alpha}}_{d/H_k}^+,\ \hat{\boldsymbol{q}}_{c/H_k}^+=\hat{\boldsymbol{q}}_{c/H_k}^-+\frac{1}{2}\boldsymbol{\Xi}(\hat{\boldsymbol{q}}_{c/H_k}^-)\delta\ \hat{\boldsymbol{\alpha}}_{c/H_k}^+$ $\hat{\boldsymbol{\beta}}_{d_k}^+=\hat{\boldsymbol{\beta}}_{d_k}^-+\Delta\hat{\boldsymbol{\beta}}_{d_k}^+,\ \hat{\boldsymbol{\omega}}_{c/I_k}^{c+}=\hat{\boldsymbol{\omega}}_{c/I_k}^{c-}+\Delta\hat{\boldsymbol{\omega}}_{c/I_k}^{c+},\ \hat{\boldsymbol{l}}_k^+=\hat{\boldsymbol{l}}_k^-+\Delta\hat{\boldsymbol{l}}_k^+,\ \hat{\boldsymbol{x}}_{p_k}^+=\hat{\boldsymbol{x}}_{p_k}^-+\Delta\hat{\boldsymbol{x}}_{p_k}^+$
传播	$\hat{\boldsymbol{\omega}}_{d/I}^d=\tilde{\boldsymbol{\omega}}_{d/I}^d-\hat{\boldsymbol{\beta}}_d,\ \hat{\boldsymbol{\omega}}_{c/I}^c=\int\boldsymbol{\psi}(\hat{\boldsymbol{\omega}}_{c/I}^c,\hat{\boldsymbol{l}})\mathrm{d}t,\ \hat{\boldsymbol{x}}_p=\int\boldsymbol{f}(\hat{\boldsymbol{x}}_p)\mathrm{d}t$ $\dot{\hat{\boldsymbol{q}}}_{d/H}=\frac{1}{2}\Xi(\hat{\boldsymbol{q}}_{d/H})\hat{\boldsymbol{\omega}}_{d/H}^d,\ \dot{\hat{\boldsymbol{q}}}_{c/H}=\frac{1}{2}\Xi(\hat{\boldsymbol{q}}_{c/H})\hat{\boldsymbol{\omega}}_{c/H}^c,\ \boldsymbol{P}_{k+1}^-=\boldsymbol{\Phi}_k\boldsymbol{P}_k^+\boldsymbol{\Phi}_k^{\mathrm{T}}+\boldsymbol{Q}_k$

4.3.3 仿真实验与分析

本节针对主星失控翻滚这一情形对所设计的 EKF 导航滤波器进行数值仿真验证。仿真参数设置如下：主星的主惯量分别是 $J_{cxx}=8\mathrm{kg}\cdot\mathrm{m}^2$，$J_{cyy}=5\mathrm{kg}\cdot\mathrm{m}^2$ 和 $J_{czz}=4\mathrm{kg}\cdot\mathrm{m}^2$，惯量比真值分别是 $l_x=1/8$，$l_y=-4/5$ 和 $l_z=3/4$；主星的初始姿态角速度为 $\boldsymbol{\omega}_{c/I}^c(t_0)=[0,0.0011,-0.0011]^{\mathrm{T}}\mathrm{rad/s}$，真实姿态运动由式（4.86）描述；从星的角速度始终为 $\boldsymbol{\omega}_{d/I}^d=[-0.002,0,0.0011]^{\mathrm{T}}\mathrm{rad/s}$。表 4.6 列出了其余仿真参数真值，主星上 6 个标志点的位置坐标见表 4.4。

表 4.6 仿真参数

仿真参数	数 值
从星陀螺参数	$\boldsymbol{\beta}_d=[1,1,1]^{\mathrm{T}}((°)/\mathrm{h})$ $\sigma_{dv}=\sqrt{10}\times10^{-5}\mathrm{rad/s^{1/2}},\sigma_{du}=\sqrt{10}\times10^{-10}\mathrm{rad/s^{3/2}}$
VISNAV 测量噪声	$\sigma_i=1.8'',\ i=1,2,\cdots,6$
扰动加速度过程噪声	$\sigma_{\varpi}=\sqrt{10}\times10^{-11}\mathrm{m/s^{3/2}}$
主星扰动力矩	$\sigma_{d_c'}=1\times10^{-5}\mathrm{N}\cdot\mathrm{m}$
初始四元数	$\boldsymbol{q}_{d/H}=[0,0,0,1]^{\mathrm{T}},\boldsymbol{q}_{c/H}=[0,0,0,1]^{\mathrm{T}}$
初始相对位置	$\boldsymbol{\rho}(0)=[200,200,100]^{\mathrm{T}}\mathrm{m}$

续表

仿真参数	数　值
初始相对速度	$\dot{\boldsymbol{\rho}}(0)=[0.01,-0.4325,0.01]^{\mathrm{T}}\mathrm{m/s}$
初始主星轨道半径和径向速率	$r_c(0)=6.986417\times10^6\mathrm{m},\ \dot{r}_c(0)=0$
初始真近点角和真近点角速率	$\theta(0)=0,\ \dot{\theta}(0)=0.00108\mathrm{rad/s}$

对于相对姿态四元数 $\boldsymbol{q}_{d/H}$ 和 $\boldsymbol{q}_{c/H}$，分别加上 1σ 姿态误差；从星陀螺漂移、主星角速度和主星惯量比滤波初值均设为零；相对位置和速度滤波初值分别设为 $\hat{\boldsymbol{\rho}}(0)=0.9\boldsymbol{\rho}(0)$ 和 $\hat{\dot{\boldsymbol{\rho}}}(0)=0.9\dot{\boldsymbol{\rho}}(0)$，其余状态量滤波初值均设为真值。滤波初始方差阵为

$$\boldsymbol{P}_0=\mathrm{blkdiag}[(\pi/180)^2\boldsymbol{I}_{3\times3}(\pi/180)^2\boldsymbol{I}_{3\times3}[2\times(\pi/(180\times3600))]^2\boldsymbol{I}_{3\times3}\ 0.01^2\boldsymbol{I}_{3\times3}\ \boldsymbol{I}_{3\times3}5\boldsymbol{I}_{3\times3}\ 0.02\boldsymbol{I}_{3\times3}\ 1000\ \ 0.01\ \ 1\times10^{-4}\ 1\times10^{-4}]$$

本节将依据观测标志点数目设定不同的仿真场景，所有仿真结果均是 100 次蒙特卡洛仿真求平均的结果。这里，采用时间平均均方根误差（Root Time-Average Mean Square Error，RTAMSE）来衡量不同标志点数目场景下的滤波性能。RTAMSE 定义如下[14]：

$$\mathrm{RTAMSE}=\sqrt{\frac{1}{N}\frac{1}{K}\sum_{n=1}^{N}\sum_{k=1}^{K}\|\boldsymbol{e}_k^n\|^2}\tag{4.114}$$

式中：N 为蒙特卡洛仿真次数；K 为一次仿真中总的步长；$\boldsymbol{e}_k^n$ 为第 n 次蒙特卡洛仿真中第 k 步估计误差。相对位置和速度的 RTAMSE 值较容易计算，而相对姿态估计误差通常计算为

$$\delta\alpha_{d/c}=2\arccos(\delta q_{d/c,4})\tag{4.115}$$

式中：$\delta q_{d/c,4}$ 为误差四元数 $\delta\boldsymbol{q}_{d/c}$ 的第四分量，定义为

$$\delta\boldsymbol{q}_{d/c}\triangleq\boldsymbol{q}_{d/c}\otimes\hat{\boldsymbol{q}}_{d/c}^{-1}\tag{4.116}$$

每次仿真实验中最后 50min 数据用来计算 RTAMSE 值，相关结果如表 4.7 所示。

表 4.7　不同仿真场景下的 RTAMSE 值

标志点个数	相对位置/m	相对速度/(m/s)	相对姿态/(°)
1	39.902833	0.031501	122.057885
2	18.335906	0.019717	41.436254
3	0.052007	0.000050	0.049735

续表

标志点个数	相对位置/m	相对速度/(m/s)	相对姿态/(°)
4	0.030572	0.000028	0.041696
5	0.026233	0.000024	0.038248
6	0.027460	0.000025	0.037989

针对 VISNAV 这种视线测量敏感器，Sun 和 Crassidis[15] 做了相对位姿确定的可观测性分析，并指出仅有一或两个视线观测量，该系统是不可观的。这是因为相对平动和转动模型以及状态估计方法，提高了 VISNAV 系统的可观性。随着标志点数目的增加，相对位置和姿态估计精度也随之提高。

图 4.17~图 4.25 为观测三个标志点情形下的仿真结果。图 4.17 和图 4.18 分别给出了相对四元数 $\boldsymbol{q}_{d/H}$ 和 $\boldsymbol{q}_{c/H}$ 的姿态估计误差与 3σ 界。由图 4.19 可以看出，从星陀螺漂移得到很好的估计。图 4.20 给出了主星角速度估计误差与 3σ 界，单轴姿态角速度估计精度在 5×10^{-6}rad/s 以内。图 4.21 给出了主星惯量比估计误差与 3σ 界。经过 50min 后，惯量比估计误差可以收敛到 0.01 以内，最终惯量比各分量估计误差在 0.001 以内。图 4.22 给出了两卫星间的相对姿态估计误差，单轴姿态估计精度在 0.02°以内。图 4.23 和图 4.24 分别给出了相对位置和相对速度估计误差，估计误差始终控制在 3σ 界内。最终，三个方向的相对位置和速度误差均分别控制在 0.05m 和 5×10^{-5} m/s 以内。图 4.25 给出了主星轨道元素估计误差与 3σ 界。

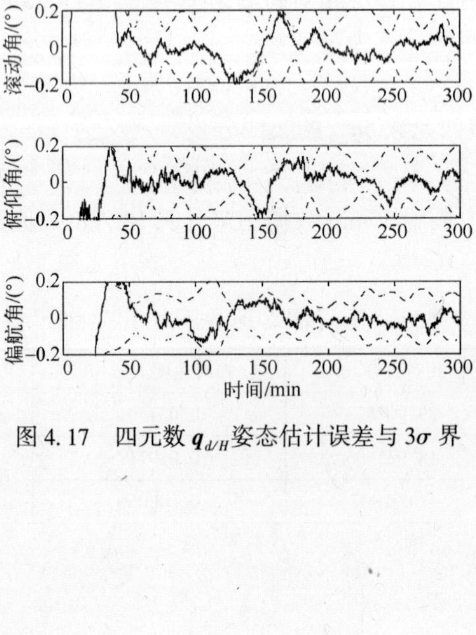

图 4.17　四元数 $\boldsymbol{q}_{d/H}$ 姿态估计误差与 3σ 界

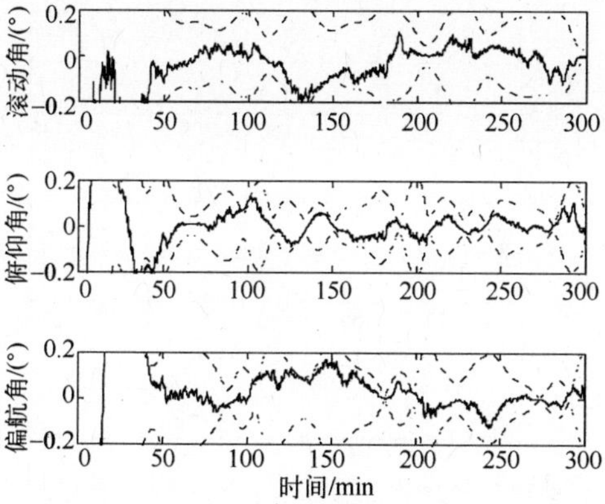

图 4.18　四元数 $\boldsymbol{q}_{c/H}$ 姿态估计误差与 3σ 界

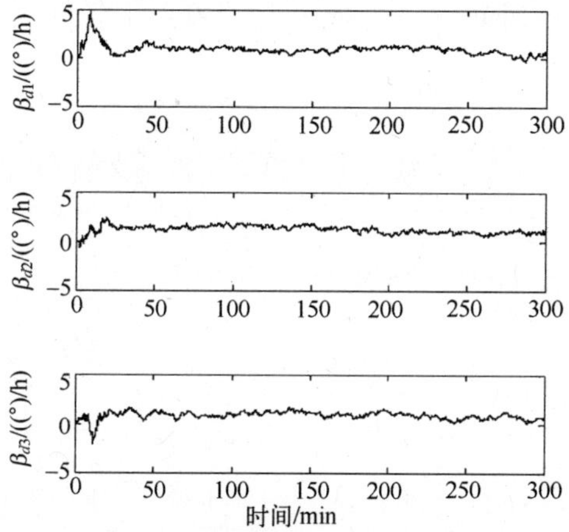

图 4.19　从星陀螺漂移估计

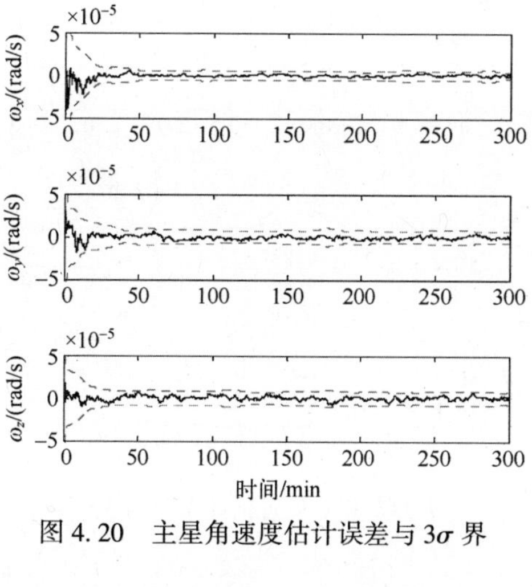

图 4.20　主星角速度估计误差与 3σ 界

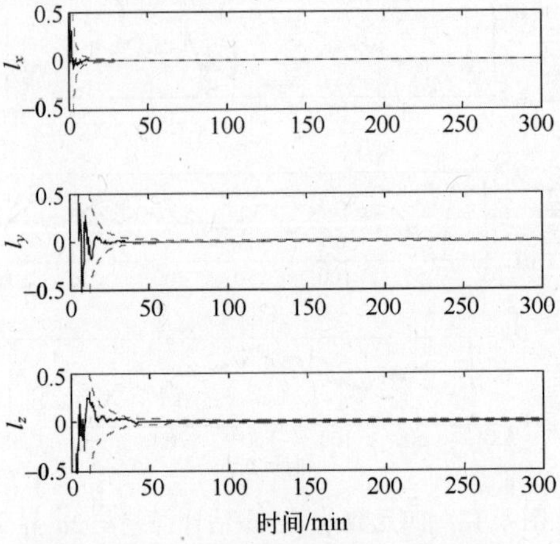

图 4.21　主星惯量比估计误差与 3σ 界

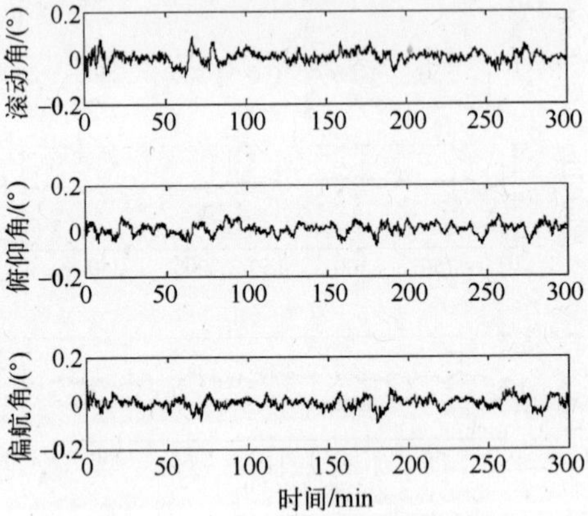

图 4.22　相对姿态估计误差

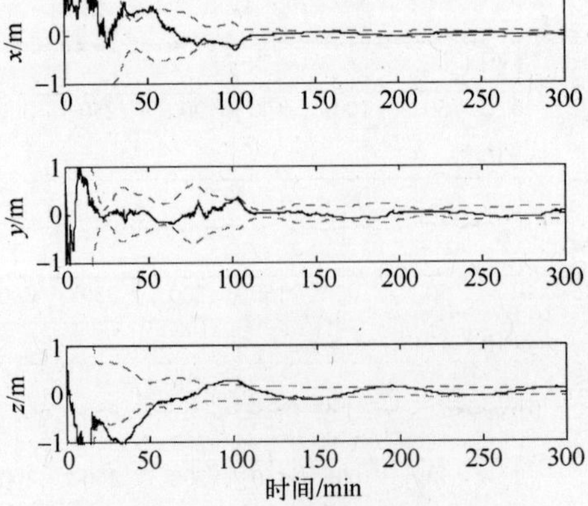

图 4.23　相对位置估计误差与 3σ 界

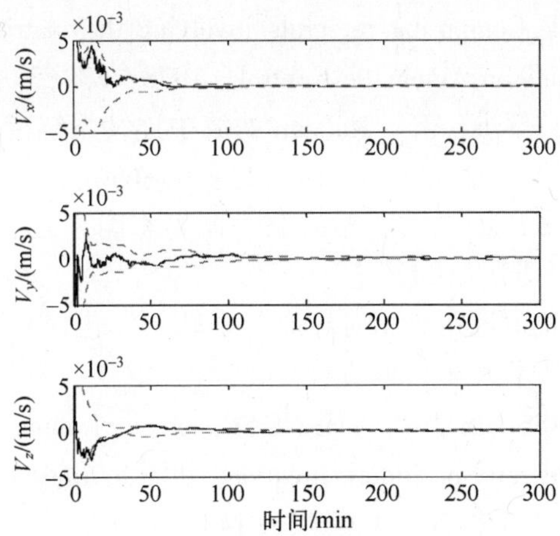

图 4.24　相对速度估计误差与 3σ 界

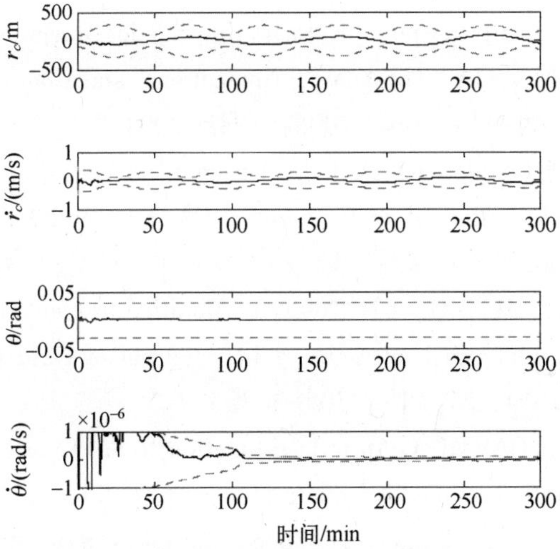

图 4.25　主星轨道元素估计误差与 3σ 界

参考文献

[1] KALMAN R E. A new approach to linear filtering and prediction problems [J]. Transactions of the ASME, Journal of Basic Engineering, 1960, 82: 34-45.

[2] MARKLEY F L. Attitude error representations for Kalman filtering [J]. Journal of Guidance, Control, and Dynamics, 2003, 26 (2): 311-317.

[3] VAN LOAN C F. Computing integrals involving the matrix exponential [J]. IEEE Transactions on Automatic Control, 1978, 23 (3): 395-404.

[4] 张力军, 钱山, 蔡洪, 等. Kalman 滤波中连续系统离散化的计算机实现 [J]. 飞行器测控学报, 2010, 29 (2): 66-69.

[5] 张力军. 基于多视场星敏感器的航天器姿态确定方法研究 [D]. 长沙: 国防科学技术大学, 2011.

[6] 张力军. 椭圆轨道近距离相对导航与姿轨一体化控制方法研究 [D]. 长沙: 国防科学技术大学, 2016.

[7] KIM S G, CRASSIDIS J L, CHENG Y, et al. Kalman filtering for relative spacecraft attitude and position estimation [J]. Journal of Guidance, Control, and Dynamics, 2007, 30 (1): 133-143.

[8] ARASARATNAM I, HAYKIN S. Cubature Kalman filters [J]. IEEE Transaction on Automatic Control, 2009, 54 (6): 1254-1269.

[9] CRASSIDIS J L. Sigma-point Kalman filtering for integrated gps and inertial navigation [C]//AIAA Guidance, Navigation, and Control Conference, San Francisco. USA: AIAA Inc, 2005: 1981-2004.

[10] CRASSIDIS J L, MARKLEY F L. Unscented filtering for spacecraft attitude estimation [J]. Journal of Guidance, Control, and Dynamics, 2003, 26 (4): 536-542.

[11] CHANG L B, HU B Q, CHANG G B. Modified unscented quaternion estimator based on quaternion averaging [J]. Journal of Guidance, Control, and Dynamics, 2014, 37 (1): 305-308.

[12] OSHMAN Y, CARMI A. Attitude estimation from vector observations using a genetic-algorithm-embedded quaternion particle filter [J]. Journal of Guidance, Control, and Dynamics, 2006, 29 (4): 879-891.

[13] MARKLEY F L, CHENG Y, CRASSIDIS J L, et al. Averaging quaternions [J]. Journal of Guidance, Control and Dynamics, 2007, 30 (4): 1193-1197.

[14] ARULAMPALAM M S, RISTIC B, GORDON N, et al. Bearings - only tracking of maneuvering targets using particle filters [J]. Eurasia Journal on Applied Signal Processing, 2004, 15: 2351-2365.

[15] SUN D B, CRASSIDIS J L. Observability analysis of six-degree-of freedom configuration determination using vector observations [J]. Journal of Guidance, Control and Dynamics, 2002, 25 (6): 1149-1157.

第 5 章　基于对偶代数的航天器姿轨一体化控制

本章将建立基于对偶代数的航天器轨/姿运动学和动力学模型，进一步推导基于对偶代数的姿轨耦合误差动力学方程，并证明其与传统的相对转动和相对平动误差动力学方程一致，但相比以往轨道和姿态分别建模的情形，形式更加简洁、明了，计算效率更高。在此基础上，深入研究基于对偶代数的航天器姿轨一体化控制方法，基于 Lyapunov 稳定性理论证明该方法的全局收敛性，并设计控制分配算法，实现全推力器姿轨一体化控制。

5.1　基于对偶代数的航天器轨/姿运动学和动力学建模

5.1.1　对偶四元数

实际上，对偶四元数是元素为对偶数的四元数，也可以写成一个元素为四元数的对偶数，即

$$\hat{\boldsymbol{q}}=\boldsymbol{q}+\varepsilon\boldsymbol{q}',\quad \varepsilon^2=0 \text{ 但 } \varepsilon\neq 0 \tag{5.1}$$

其中，$\boldsymbol{q}$ 和 $\boldsymbol{q}'$都是普通的四元数（它们的范数不一定为 1）。显然，对偶矢量可看作标量部分为零的对偶四元数。对偶四元数与普通四元数具有相同的性质，即

$$\hat{\boldsymbol{q}}_1+\hat{\boldsymbol{q}}_2=(\boldsymbol{q}_1+\boldsymbol{q}_2)+\varepsilon(\boldsymbol{q}_1'+\boldsymbol{q}_2') \tag{5.2}$$

$$\lambda\,\hat{\boldsymbol{q}}=\lambda\boldsymbol{q}+\varepsilon\lambda\boldsymbol{q}',\ \forall\,\lambda\in\mathbb{R} \tag{5.3}$$

$$\hat{\boldsymbol{q}}_1\otimes\hat{\boldsymbol{q}}_2=\boldsymbol{q}_1\otimes\boldsymbol{q}_2+\varepsilon(\boldsymbol{q}_1\otimes\boldsymbol{q}_2'+\boldsymbol{q}_1'\otimes\boldsymbol{q}_2) \tag{5.4}$$

在实际运算和编程实现时，对偶四元数常可表示成八维列矢量形式，即 $\hat{\boldsymbol{q}}=[\boldsymbol{q}^{\mathrm{T}}\boldsymbol{q}'^{\mathrm{T}}]^{\mathrm{T}}$。那么，式（5.4）又可写成矩阵形式，即

$$\begin{aligned}\hat{\boldsymbol{q}}_1\otimes\hat{\boldsymbol{q}}_2&=\begin{bmatrix}\{\boldsymbol{q}_1\}_L & \boldsymbol{0}_{4\times4}\\ \{\boldsymbol{q}_1'\}_L & \{\boldsymbol{q}_1\}_L\end{bmatrix}\begin{bmatrix}\boldsymbol{q}_2\\ \boldsymbol{q}_2'\end{bmatrix}\\ &=\begin{bmatrix}\{\boldsymbol{q}_2\}_R & \boldsymbol{0}_{4\times4}\\ \{\boldsymbol{q}_2'\}_R & \{\boldsymbol{q}_2\}_R\end{bmatrix}\begin{bmatrix}\boldsymbol{q}_1\\ \boldsymbol{q}_1'\end{bmatrix}\end{aligned} \tag{5.5}$$

式中：$\{\cdot\}_L$ 和 $\{\cdot\}_R$ 的定义分别见式（2.11）和式（2.12）。值得注意的是，这里对偶四元数乘法规则沿用了2.2节中的四元数乘法规则，与Hamiliton建立的常用四元数乘法规则恰好相反。因此，使得本章后续推导的一些基于对偶代数的公式在形式上与参考文献［1-2］中不完全一致，但它们本质上是一致的。

对偶四元数的范数定义为

$$\begin{aligned}\|\hat{\boldsymbol{q}}\| &= \sqrt{\|\hat{\boldsymbol{q}}\|^2} = \sqrt{\hat{\boldsymbol{q}} \otimes \hat{\boldsymbol{q}}^*} \\ &= \sqrt{(\boldsymbol{q}+\varepsilon\boldsymbol{q}') \otimes (\boldsymbol{q}^*+\varepsilon\boldsymbol{q}'^*)} \\ &= \sqrt{\boldsymbol{q} \otimes \boldsymbol{q}^* + \varepsilon(\boldsymbol{q} \otimes \boldsymbol{q}'^* + \boldsymbol{q}' \otimes \boldsymbol{q}^*)} \\ &= \sqrt{\boldsymbol{q}^{\mathrm{T}}\boldsymbol{q} + \varepsilon 2\boldsymbol{q}^{\mathrm{T}}\boldsymbol{q}'} \\ &= \sqrt{\boldsymbol{q}^{\mathrm{T}}\boldsymbol{q}} + \varepsilon \frac{\boldsymbol{q}^{\mathrm{T}}\boldsymbol{q}'}{\sqrt{\boldsymbol{q}^{\mathrm{T}}\boldsymbol{q}}}\end{aligned} \tag{5.6}$$

这是一个对偶数。如果对偶四元数的实数部分非零，其逆可以写作 $\hat{\boldsymbol{q}}^{-1} = \hat{\boldsymbol{q}}^* / \|\hat{\boldsymbol{q}}\|^2$，其中 $\hat{\boldsymbol{q}}^* = \boldsymbol{q}^* + \varepsilon\boldsymbol{q}'^*$ 是共轭对偶四元数。若 $\|\hat{\boldsymbol{q}}\|^2 = 1+\varepsilon 0$，则称该对偶四元数为单位对偶四元数，因此，单位对偶四元数的逆存在且等于其共轭。

单位对偶四元数可以用来描述包括旋转和平移的一般刚体变换，即螺旋变换。假定坐标系 M 与坐标系 N 之间的一般性刚体运动可由转动 $\boldsymbol{q}$ 紧接着平移 $\boldsymbol{t}^N$（或平移 $\boldsymbol{t}^M$ 紧接着转动 $\boldsymbol{q}$）描述。假设Plücker直线 $\hat{\boldsymbol{L}}$ 在坐标系 M 和 N 中分别表示为 $\hat{\boldsymbol{L}}^M = \boldsymbol{l}^M + \varepsilon\boldsymbol{m}^M$ 和 $\hat{\boldsymbol{L}}^N = \boldsymbol{l}^N + \varepsilon\boldsymbol{m}^N$，那么可定义一个单位对偶四元数 $\hat{\boldsymbol{q}}$ 使其满足 $\hat{\boldsymbol{L}}^N = \hat{\boldsymbol{q}} \otimes \hat{\boldsymbol{L}}^M \otimes \hat{\boldsymbol{q}}^{-1}$，其中单位对偶四元数 $\hat{\boldsymbol{q}}$ 是关于 $\boldsymbol{q}$ 和 $\boldsymbol{t}^N$（或 $\boldsymbol{q}$ 和 $\boldsymbol{t}^M$）的函数，即

$$\begin{aligned}\hat{\boldsymbol{q}} &= \boldsymbol{q} + \varepsilon\boldsymbol{q}' \\ &= \boldsymbol{q} + \varepsilon \frac{1}{2}\boldsymbol{t}^N \otimes \boldsymbol{q} = \boldsymbol{q} + \varepsilon \frac{1}{2}\boldsymbol{q} \otimes \boldsymbol{t}^M\end{aligned} \tag{5.7}$$

需要说明的是，当遇到运算符号“⊗”时，本章中的三维矢量和对偶矢量自动扩展为相应的广义四元数以及对偶四元数进行运算，不另做符号标记。

由前述知识可知，单位对偶四元数的逆存在且等于其共轭，即

$$\begin{aligned}\hat{\boldsymbol{q}}^{-1} &= \hat{\boldsymbol{q}}^* = \boldsymbol{q}^* + \varepsilon\boldsymbol{q}'^* \\ &= \boldsymbol{q}^* - \varepsilon \frac{1}{2}\boldsymbol{q}^* \otimes \boldsymbol{t}^N = \boldsymbol{q}^* - \varepsilon \frac{1}{2}\boldsymbol{t}^M \otimes \boldsymbol{q}^*\end{aligned} \tag{5.8}$$

根据参考文献［2］，可知对偶四元数的运动学方程为

$$\dot{\hat{\boldsymbol{q}}} = \frac{1}{2}\hat{\boldsymbol{\omega}}_{N/M}^N \otimes \hat{\boldsymbol{q}} = \frac{1}{2}\hat{\boldsymbol{q}} \otimes \hat{\boldsymbol{\omega}}_{N/M}^M \tag{5.9}$$

式中：对偶矢量

$$\hat{\boldsymbol{\omega}}_{N/M}^{N}=\boldsymbol{\omega}_{N/M}^{N}+\varepsilon(\dot{\boldsymbol{t}}^{N}+\boldsymbol{\omega}_{N/M}^{N}\times\boldsymbol{t}^{N}) \tag{5.10}$$

$$\hat{\boldsymbol{\omega}}_{N/M}^{M}=\boldsymbol{\omega}_{N/M}^{M}+\varepsilon(\dot{\boldsymbol{t}}^{M}+\boldsymbol{t}^{M}\times\boldsymbol{\omega}_{N/M}^{M}) \tag{5.11}$$

被称为旋量。这里，$\boldsymbol{\omega}_{N/M}^{N}$和 $\boldsymbol{\omega}_{N/M}^{M}$分别为坐标系 N 相对于坐标系 M 的角速度在 N 系和 M 系中的分量。值得注意的是，本章采用的四元数乘法规则与参考文献［2］中定义的四元数乘法规则恰好相反，因此式（5.7）和式（5.9）在形式上与参考文献［2］中稍有区别，但它们本质上是一致的。另外，经推导，式（5.10）和式（5.11）中的旋量形式与参考文献［2］中保持一致。

5.1.2　基于对偶代数的航天器轨/姿运动学方程

如图 5.1 所示，假设对偶四元数$\hat{\boldsymbol{q}}_d$ 刻画了从星本体坐标系（d 系）相对于地心惯性坐标系（I 系）的一般性刚体运动，由定义可知[3]

$$\begin{aligned}\hat{\boldsymbol{q}}_d&=\boldsymbol{q}_d+\varepsilon\frac{1}{2}\boldsymbol{q}_d\otimes\boldsymbol{r}_d^{I}\\&=\boldsymbol{q}_d+\varepsilon\frac{1}{2}\boldsymbol{r}_d^{d}\otimes\boldsymbol{q}_d\end{aligned} \tag{5.12}$$

式中：$\boldsymbol{q}_d$ 为从星体系相对于惯性系的姿态四元数；$\boldsymbol{r}_d^{I}$ 和 $\boldsymbol{r}_d^{d}$ 分别为从星的位置矢量在地心惯性系 I 系和从星体系 d 系中的描述。

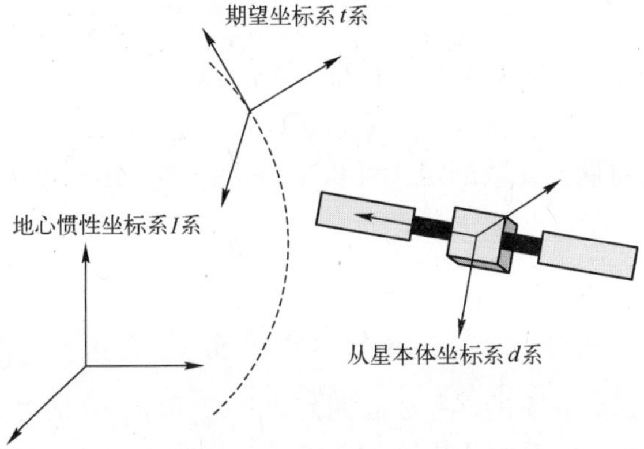

图 5.1　坐标系定义

根据前述的知识可得基于对偶代数的从星轨/姿运动学方程为

$$\dot{\hat{\boldsymbol{q}}}_d=\frac{1}{2}\hat{\boldsymbol{\omega}}_{d/I}^{d}\otimes\hat{\boldsymbol{q}}_d \tag{5.13}$$

式中：旋量$\hat{\boldsymbol{\omega}}_{d/I}^{d}$为

$$\begin{aligned}\hat{\boldsymbol{\omega}}_{d/I}^{d}&=\boldsymbol{\omega}_{d/I}^{d}+\varepsilon(\dot{\boldsymbol{r}}_{d}^{d}+\boldsymbol{\omega}_{d/I}^{d}\times\boldsymbol{r}_{d}^{d})\\&=\boldsymbol{\omega}_{d/I}^{d}+\varepsilon(\boldsymbol{q}_{d}\otimes\dot{\boldsymbol{r}}_{d}^{I}\otimes\boldsymbol{q}_{d}^{*})\\&=\boldsymbol{\omega}_{d/I}^{d}+\varepsilon\boldsymbol{v}_{d}^{d}\end{aligned}\tag{5.14}$$

式中：$\boldsymbol{\omega}_{d/I}^{d}$为从星体系相对于惯性系的角速度在航天器体系中的分量；$\boldsymbol{v}_{d}^{d}$为从星的惯性速度矢量在本体系中的描述。

同理，可用对偶四元数$\hat{\boldsymbol{q}}_{t}$描述定义在期望坐标系 t 系中的期望姿态和位置运动，即

$$\begin{aligned}\hat{\boldsymbol{q}}_{t}&=\boldsymbol{q}_{t}+\varepsilon\frac{1}{2}\boldsymbol{q}_{t}\otimes\boldsymbol{r}_{t}^{I}\\&=\boldsymbol{q}_{t}+\varepsilon\frac{1}{2}\boldsymbol{r}_{t}^{t}\otimes\boldsymbol{q}_{t}\end{aligned}\tag{5.15}$$

式中：$\boldsymbol{q}_{t}$为从 I 系到 t 系的姿态四元数，即期望姿态；$\boldsymbol{r}_{t}^{I}$和$\boldsymbol{r}_{t}^{t}$分别为期望的位置矢量在 I 系和 t 系中的描述。期望坐标系 t 系的运动学方程为

$$\dot{\hat{\boldsymbol{q}}}_{t}=\frac{1}{2}\hat{\boldsymbol{\omega}}_{t/I}^{t}\otimes\hat{\boldsymbol{q}}_{t}\tag{5.16}$$

其中，$\hat{\boldsymbol{\omega}}_{t/I}^{t}$为期望旋量，类似于式（5.14）可写成

$$\hat{\boldsymbol{\omega}}_{t/I}^{t}=\boldsymbol{\omega}_{t/I}^{t}+\varepsilon(\dot{\boldsymbol{r}}_{t}^{t}+\boldsymbol{\omega}_{t/I}^{t}\times\boldsymbol{r}_{t}^{t})=\boldsymbol{\omega}_{t/I}^{t}+\varepsilon\boldsymbol{v}_{t}^{t}\tag{5.17}$$

式中：$\boldsymbol{\omega}_{t/I}^{t}$，$\boldsymbol{r}_{t}^{t}$和$\boldsymbol{v}_{t}^{t}$分别为定义在 t 系中的期望惯性角速度矢量、惯性位置矢量以及速度矢量。

定义 d 系和 t 系之间的误差对偶四元数$\hat{\boldsymbol{q}}_{e}$为

$$\hat{\boldsymbol{q}}_{e}=\hat{\boldsymbol{q}}_{d/t}=\hat{\boldsymbol{q}}_{d}\otimes\hat{\boldsymbol{q}}_{t}^{-1}=\hat{\boldsymbol{q}}_{d}\otimes\hat{\boldsymbol{q}}_{t}^{*}\tag{5.18}$$

一方面，由对偶四元数的定义可知

$$\begin{aligned}\hat{\boldsymbol{q}}_{e}&=\boldsymbol{q}_{e}+\varepsilon\frac{1}{2}\boldsymbol{q}_{e}\otimes\boldsymbol{p}_{d/t}^{t}\\&=\boldsymbol{q}_{e}+\varepsilon\frac{1}{2}\boldsymbol{p}_{d/t}^{d}\otimes\boldsymbol{q}_{e}\end{aligned}\tag{5.19}$$

式中：$\boldsymbol{q}_{e}$为从 t 系到 d 系的误差姿态四元数；$\boldsymbol{p}_{d/t}^{t}$和$\boldsymbol{p}_{d/t}^{d}$分别为 d 系相对于 t 系的位置矢量在 t 系和 d 系中的描述。它们可表示为

$$\boldsymbol{q}_{e}=\boldsymbol{q}_{d}\otimes\boldsymbol{q}_{t}^{-1}=\boldsymbol{q}_{d}\otimes\boldsymbol{q}_{t}^{*}\tag{5.20}$$

$$\boldsymbol{p}_{d/t}^{t}=\boldsymbol{q}_{e}^{*}\otimes\boldsymbol{r}_{d}^{d}\otimes\boldsymbol{q}_{e}-\boldsymbol{r}_{t}^{t}=\boldsymbol{r}_{d}^{t}-\boldsymbol{r}_{t}^{t}\tag{5.21}$$

$$\boldsymbol{p}_{d/t}^{d}=\boldsymbol{r}_{d}^{d}-\boldsymbol{q}_{e}\otimes\boldsymbol{r}_{t}^{t}\otimes\boldsymbol{q}_{e}^{*}=\boldsymbol{r}_{d}^{d}-\boldsymbol{r}_{t}^{d}\tag{5.22}$$

另一方面，将式（5.12）和式（5.15）代入式（5.18）也可得

$$\begin{aligned}\hat{\boldsymbol{q}}_e &= \left(\boldsymbol{q}_d+\varepsilon\frac{1}{2}\boldsymbol{r}_d^d\otimes\boldsymbol{q}_d\right)\otimes\left(\boldsymbol{q}_t^*-\varepsilon\frac{1}{2}\boldsymbol{q}_t^*\otimes\boldsymbol{r}_t^t\right)\\ &=\boldsymbol{q}_e+\varepsilon\frac{1}{2}(\boldsymbol{r}_d^d-\boldsymbol{q}_e\otimes\boldsymbol{r}_t^t\otimes\boldsymbol{q}_e^*)\otimes\boldsymbol{q}_e\\ &=\boldsymbol{q}_e+\varepsilon\frac{1}{2}\boldsymbol{p}_{d/t}^d\otimes\boldsymbol{q}_e\end{aligned}\tag{5.23}$$

这也从侧面验证了式（5.19）的正确性。

依据定义，可得误差对偶四元数的运动学方程为[4]

$$\dot{\hat{\boldsymbol{q}}}_e=\frac{1}{2}\hat{\boldsymbol{\omega}}_{d/t}^d\otimes\hat{\boldsymbol{q}}_e\tag{5.24}$$

式中：d 系相对于 t 系的误差旋量$\hat{\boldsymbol{\omega}}_{d/t}^d$为

$$\hat{\boldsymbol{\omega}}_{d/t}^d=\boldsymbol{\omega}_{d/t}^d+\varepsilon(\dot{\boldsymbol{p}}_{d/t}^d+\boldsymbol{\omega}_{d/t}^d\times\boldsymbol{p}_{d/t}^d)\tag{5.25}$$

其中，$\boldsymbol{\omega}_{d/t}^d$为 d 系相对于 t 系的角速度在 d 系中的分量。

另外，误差旋量$\hat{\boldsymbol{\omega}}_{d/t}^d$也可以通过下式计算得到[5]

$$\begin{aligned}\hat{\boldsymbol{\omega}}_{d/t}^d&=\hat{\boldsymbol{\omega}}_{d/I}^d-\hat{\boldsymbol{q}}_e\otimes\hat{\boldsymbol{\omega}}_{t/I}^t\otimes\hat{\boldsymbol{q}}_e^*\\ &=\boldsymbol{\omega}_{d/t}^d+\varepsilon(\dot{\boldsymbol{p}}_{d/t}^d+\boldsymbol{\omega}_{d/t}^d\times\boldsymbol{p}_{d/t}^d)\end{aligned}\tag{5.26}$$

详细推导见附录 C。

5.1.3　基于对偶代数的轨/姿一体化动力学方程

根据参考文献［6-7］中的 Brodsky 理论，刚体的对偶动量可以表示为

$$\hat{\boldsymbol{h}}=\hat{\boldsymbol{M}}\hat{\boldsymbol{\omega}}\tag{5.27}$$

其中，$\hat{\boldsymbol{\omega}}$为刚体的旋量；$\hat{\boldsymbol{M}}$为对偶质量算子和对偶惯量算子的和，定义为对偶惯性算子，其矩阵形式为

$$\begin{aligned}\hat{\boldsymbol{M}}&=\left(m\frac{d}{d\varepsilon}\right)\boldsymbol{I}_{3\times3}+\varepsilon\boldsymbol{J}\\ &=\begin{bmatrix}m\frac{d}{d\varepsilon}+\varepsilon J_{xx} & \varepsilon J_{xy} & \varepsilon J_{xz}\\ \varepsilon J_{xy} & m\frac{d}{d\varepsilon}+\varepsilon J_{yy} & \varepsilon J_{yz}\\ \varepsilon J_{xz} & \varepsilon J_{yz} & m\frac{d}{d\varepsilon}+\varepsilon J_{zz}\end{bmatrix}\end{aligned}\tag{5.28}$$

式中：m 为航天器质量；$\boldsymbol{J}$ 为航天器的惯量张量；$\boldsymbol{I}_{3\times3}$为单位阵，且有$\hat{\boldsymbol{M}}^{-1}=\boldsymbol{J}^{-1}\frac{d}{d\varepsilon}+\varepsilon\frac{1}{m}\boldsymbol{I}_{3\times3}$。

因此，对偶动量可进一步写成

$$\hat{\boldsymbol{h}}=\widehat{\boldsymbol{M}}\hat{\boldsymbol{\omega}}=\left(\left(m\frac{d}{d\varepsilon}\right)\boldsymbol{I}_{3\times3}+\varepsilon\boldsymbol{J}\right)(\boldsymbol{\omega}+\varepsilon\boldsymbol{v})=m\boldsymbol{v}+\varepsilon\boldsymbol{J}\boldsymbol{\omega} \tag{5.29}$$

可以看出，对偶动量$\hat{\boldsymbol{h}}$的实部 $m\boldsymbol{v}$ 和对偶部 $\boldsymbol{J}\boldsymbol{\omega}$ 分别为刚体的线动量和角动量。如果以体坐标系为计算坐标系，根据动量矩定理可得

$$\frac{\partial\hat{\boldsymbol{h}}}{\partial t}+\hat{\boldsymbol{\omega}}\times\hat{\boldsymbol{h}}=\hat{\boldsymbol{F}} \tag{5.30}$$

式中：$\hat{\boldsymbol{F}}$为作用在刚体上的对偶力矢量，由力 $\boldsymbol{F}$ 和力矩 $\boldsymbol{T}$ 构成，即

$$\hat{\boldsymbol{F}}=\boldsymbol{F}+\varepsilon\boldsymbol{T} \tag{5.31}$$

其中，力 $\boldsymbol{F}$ 和力矩 $\boldsymbol{T}$ 都定义在刚体本体系中。若将式（5.30）进一步展开分为实部和对偶部，可得

$$\begin{cases}m\dot{\boldsymbol{v}}+m\boldsymbol{\omega}\times\boldsymbol{v}=\boldsymbol{F}\\ \boldsymbol{J}\dot{\boldsymbol{\omega}}+\boldsymbol{\omega}\times\boldsymbol{J}\boldsymbol{\omega}=\boldsymbol{T}\end{cases} \tag{5.32}$$

显然，上述两组方程分别描述了刚体随质心平动和绕质心转动两部分运动，且相互耦合。

根据式（5.30），可得基于对偶代数的从星轨/姿动力学方程为

$$\widehat{\boldsymbol{M}}_d\,\dot{\hat{\boldsymbol{\omega}}}_{d/I}^d+\hat{\boldsymbol{\omega}}_{d/I}^d\times\widehat{\boldsymbol{M}}_d\,\hat{\boldsymbol{\omega}}_{d/I}^d=\hat{\boldsymbol{F}}^d \tag{5.33}$$

式中：对偶力矢量$\hat{\boldsymbol{F}}^d$ 由作用在从星上的力 $\boldsymbol{F}^d$ 和力矩 $\boldsymbol{T}^d$ 组成。其中，力矢量 $\boldsymbol{F}^d$ 主要由控制力 $\boldsymbol{f}_c^d$、重力 $\boldsymbol{f}_g^d$ 以及干扰力 $\boldsymbol{f}_d$ 三部分构成；力矩矢量 $\boldsymbol{T}^d$ 由控制力矩 $\boldsymbol{\tau}_c^d$、重力梯度力矩 $\boldsymbol{\tau}_g^d$ 以及干扰力矩 $\boldsymbol{\tau}_d$ 三部分构成。因此，对偶力矢量$\hat{\boldsymbol{F}}^d$ 又可写成

$$\hat{\boldsymbol{F}}^d=\hat{\boldsymbol{u}}_c^d+\hat{\boldsymbol{u}}_g^d+\hat{\boldsymbol{d}} \tag{5.34}$$

其中，

$$\begin{aligned}\hat{\boldsymbol{u}}_c^d&=\boldsymbol{f}_c^d+\varepsilon\boldsymbol{\tau}_c^d\\ \hat{\boldsymbol{u}}_g^d&=\boldsymbol{f}_g^d+\varepsilon\boldsymbol{\tau}_g^d\\ \hat{\boldsymbol{d}}&=\boldsymbol{f}_d+\varepsilon\boldsymbol{\tau}_d\end{aligned} \tag{5.35}$$

重力和重力梯度力矩分别计算如下[8-9]：

$$\boldsymbol{f}_g^d=-\frac{\mu m_d}{r_d^3}\boldsymbol{r}_d^d,\quad \boldsymbol{\tau}_g^d=\frac{3\mu}{r_d^5}(\boldsymbol{r}_d^d\times\boldsymbol{J}_d\boldsymbol{r}_d^d) \tag{5.36}$$

式中：μ 为引力常数；r_d 为从星的地心距；m_d 和 $\boldsymbol{J}_d$ 分别为从星的质量和惯量张量。

结合式（5.33）和式（5.34），可得基于对偶代数的从星轨/姿一体化动力学方程

$$\widehat{\boldsymbol{M}}_d \dot{\widehat{\boldsymbol{\omega}}}_{d/I}^d+\widehat{\boldsymbol{\omega}}_{d/I}^d\times\widehat{\boldsymbol{M}}_d \widehat{\boldsymbol{\omega}}_{d/I}^d=\widehat{\boldsymbol{u}}_c^d+\widehat{\boldsymbol{u}}_g^d+\widehat{\boldsymbol{d}} \tag{5.37}$$

分析可知，从星的轨道运动和姿态运动均可以由式（5.37）传播得到，形式相当简洁，且与欧拉姿态动力学方程在形式上相似。因此，在设计相应的控制器时可借鉴已有的姿态运动控制律设计的方法。

5.1.4　误差动力学方程

近距离相对轨道和姿态运动的典型控制任务包括空间拦截、空间交会、空间悬停与空间绕飞等任务。本节在前文已推导的空间单航天器轨/姿一体化动力学方程的基础上，进一步推导基于对偶代数的航天器相对轨道和姿态误差动力学方程。

对$\widehat{\boldsymbol{q}}_e\otimes\widehat{\boldsymbol{q}}_e^*=[0\ 0\ 0\ 1]^{\mathrm{T}}+\varepsilon[0\ 0\ 0\ 0]^{\mathrm{T}}$两边求导可得

$$\dot{\widehat{\boldsymbol{q}}}_e^*=-\frac{1}{2}\widehat{\boldsymbol{q}}_e^*\otimes\widehat{\boldsymbol{\omega}}_{d/t}^d \tag{5.38}$$

进一步，对式（5.26）两边求导可得

$$\begin{aligned}
\dot{\widehat{\boldsymbol{\omega}}}_{d/t}^d&=\dot{\widehat{\boldsymbol{\omega}}}_{d/I}^d-\dot{\widehat{\boldsymbol{q}}}_e\otimes\widehat{\boldsymbol{\omega}}_{t/I}^t\otimes\widehat{\boldsymbol{q}}_e^*-\widehat{\boldsymbol{q}}_e\otimes\dot{\widehat{\boldsymbol{\omega}}}_{t/I}^t\otimes\widehat{\boldsymbol{q}}_e^*-\widehat{\boldsymbol{q}}_e\otimes\widehat{\boldsymbol{\omega}}_{t/I}^t\otimes\dot{\widehat{\boldsymbol{q}}}_e^*\\
&=\dot{\widehat{\boldsymbol{\omega}}}_{d/I}^d-\widehat{\boldsymbol{q}}_e\otimes\dot{\widehat{\boldsymbol{\omega}}}_{t/I}^t\otimes\widehat{\boldsymbol{q}}_e^*-\frac{1}{2}\widehat{\boldsymbol{\omega}}_{d/t}^d\otimes\widehat{\boldsymbol{q}}_e\otimes\widehat{\boldsymbol{\omega}}_{t/I}^t\otimes\widehat{\boldsymbol{q}}_e^*+\frac{1}{2}\widehat{\boldsymbol{q}}_e\otimes\widehat{\boldsymbol{\omega}}_{t/I}^t\otimes\widehat{\boldsymbol{q}}_e^*\otimes\widehat{\boldsymbol{\omega}}_{d/t}^d\\
&=\dot{\widehat{\boldsymbol{\omega}}}_{d/I}^d-\widehat{\boldsymbol{q}}_e\otimes\dot{\widehat{\boldsymbol{\omega}}}_{t/I}^t\otimes\widehat{\boldsymbol{q}}_e^*+\widehat{\boldsymbol{\omega}}_{d/t}^d\times(\widehat{\boldsymbol{q}}_e\otimes\widehat{\boldsymbol{\omega}}_{t/I}^t\otimes\widehat{\boldsymbol{q}}_e^*)
\end{aligned} \tag{5.39}$$

将式（5.37）代入式（5.39）可得误差动力学方程

$$\dot{\widehat{\boldsymbol{\omega}}}_{d/t}^d=-\widehat{\boldsymbol{M}}_d^{-1}(\widehat{\boldsymbol{\omega}}_{d/I}^d\times\widehat{\boldsymbol{M}}_d\widehat{\boldsymbol{\omega}}_{d/I}^d)+\widehat{\boldsymbol{M}}_d^{-1}(\widehat{\boldsymbol{u}}_c^d+\widehat{\boldsymbol{u}}_g^d+\widehat{\boldsymbol{d}})-\widehat{\boldsymbol{q}}_e\otimes\dot{\widehat{\boldsymbol{\omega}}}_{t/I}^t\otimes\widehat{\boldsymbol{q}}_e^*+\widehat{\boldsymbol{\omega}}_{d/t}^d\times(\widehat{\boldsymbol{q}}_e\otimes\widehat{\boldsymbol{\omega}}_{t/I}^t\otimes\widehat{\boldsymbol{q}}_e^*) \tag{5.40}$$

正如式（5.40）所示，从星体系 d 系相对于期望坐标系 t 系的相对平动和转动动力学方程被描述在同一个方程里，形式相当简洁。若将式（5.40）进一步展开分为实部和对偶部，可分别得到相对转动误差动力学方程和相对平动误差动力学方程，即

$$\dot{\boldsymbol{\omega}}_{d/t}^d=-\boldsymbol{J}_d^{-1}(\boldsymbol{\omega}_{d/I}^d\times\boldsymbol{J}_d\boldsymbol{\omega}_{d/I}^d)-(\boldsymbol{A}(\boldsymbol{q}_e)\dot{\boldsymbol{\omega}}_{t/I}^t-\boldsymbol{\omega}_{d/t}^d\times\boldsymbol{A}(\boldsymbol{q}_e)\boldsymbol{\omega}_{t/I}^t)+\boldsymbol{J}_d^{-1}(\boldsymbol{\tau}_c^d+\boldsymbol{\tau}_g^d+\boldsymbol{\tau}_d) \tag{5.41}$$

$$\ddot{\boldsymbol{p}}_{d/t}^d=-\frac{\mu}{r_d^3}\boldsymbol{r}_d^d+\frac{\mu}{r_t^3}\boldsymbol{A}(\boldsymbol{q}_e)\boldsymbol{r}_t^t-2\boldsymbol{\omega}_{d/I}^d\times\dot{\boldsymbol{p}}_{d/t}^d-\dot{\boldsymbol{\omega}}_{d/I}^d\times\boldsymbol{p}_{d/t}^d-\boldsymbol{\omega}_{d/I}^d\times(\boldsymbol{\omega}_{d/I}^d\times\boldsymbol{p}_{d/t}^d)+\frac{\boldsymbol{f}_c^d}{m_d}+\frac{\boldsymbol{f}_d}{m_d} \tag{5.42}$$

具体推导详见附录 C。这里，$\boldsymbol{A}(\boldsymbol{q}_e)$为从 t 系到 d 系的姿态误差阵。由

式（5.41）可以得到相对转动误差动力学方程，其形式与传统的姿态跟踪误差动力学方程一致；由式（5.42）可以得到相对平动误差动力学方程，其形式相当于将传统的相对轨道动力学方程投影到从星本体系 d 系中。因此，基于对偶代数的姿轨耦合误差动力学方程与传统的相对转动和相对平动误差动力学方程一致，但相比以往轨道和姿态分别建模的情形，式（5.40）形式更加简洁、明了，计算效率更高。

分析可知，相对轨道运动受到相对姿态运动的影响，包括姿态误差阵 $\boldsymbol{A}(\boldsymbol{q}_e)$ 和相对旋转角速度 $\boldsymbol{\omega}_{d/t}^d$ 两部分影响，而且重力梯度力矩也会引起相对位置和姿态的耦合。另外，式（5.9）中的对偶四元数运动学方程有两种表现形式，因此，从误差旋量 $\widehat{\boldsymbol{\omega}}_{d/t}^t$ 出发，可推导在期望坐标系 t 系中描述的相对平动误差动力学方程，即

$$\ddot{\boldsymbol{p}}_{d/t}^t=-\frac{\mu}{r_d^3}\boldsymbol{A}^{\mathrm{T}}(\boldsymbol{q}_e)\boldsymbol{r}_d^d+\frac{\mu}{r_t^3}\boldsymbol{r}_t^t-2\boldsymbol{\omega}_{t/I}^t\times\dot{\boldsymbol{p}}_{d/t}^t-\dot{\boldsymbol{\omega}}_{t/I}^t\times\boldsymbol{p}_{d/t}^t-\boldsymbol{\omega}_{t/I}^t\times(\boldsymbol{\omega}_{t/I}^t\times\boldsymbol{p}_{d/t}^t)+\frac{\boldsymbol{f}_c^t}{m_d}+\frac{\boldsymbol{f}_d}{m_d} \tag{5.43}$$

可以看出，若期望坐标系 t 系是主星的当地轨道坐标系，则方程（5.43）可看作传统的在 LVLH 系中描述的相对动力学方程。虽然旋转坐标角速度 $\boldsymbol{\omega}_{t/I}^t$ 或许精确已知，但其求解的控制力需要描述在 t 系中，而发动机推力是在从星体系 d 系中计算产生的，因此两者之间仍然需要考虑姿态转换，这也就是姿轨耦合问题中的控制指令耦合。

5.2 基于对偶代数的航天器姿轨一体化控制系统设计

5.2.1 系统模型不确定性

考虑到航天器质量特性参数的不确定性，从星质量 m_d 和惯量张量 $\boldsymbol{J}_d$ 可表示为

$$m_d=\overline{m}_d+\Delta m,\quad \boldsymbol{J}_d=\overline{\boldsymbol{J}}_d+\Delta\boldsymbol{J} \tag{5.44}$$

式中：$\overline{m}_d$ 和 $\overline{\boldsymbol{J}}_d$ 分别为从星的标称质量和惯量张量；Δm 和 $\Delta\boldsymbol{J}$ 为对应的不确定性量，满足

$$\|\Delta m\|\leqslant b_1,\quad \|\Delta\boldsymbol{J}\|\leqslant b_2 \tag{5.45}$$

因此，对偶惯性算子 $\widehat{\boldsymbol{M}}_d$ 也可以描述成

$$\widehat{\boldsymbol{M}}_d=\widehat{\overline{\boldsymbol{M}}}_d+\Delta\widehat{\boldsymbol{M}} \tag{5.46}$$

其中，

$$\widehat{\boldsymbol{M}}_d=\left(\overline{m}_d\frac{d}{d\varepsilon}\right)\boldsymbol{I}_{3\times3}+\varepsilon\overline{\boldsymbol{J}}_d,\Delta\widehat{\boldsymbol{M}}=\left(\Delta m\frac{d}{d\varepsilon}\right)\boldsymbol{I}_{3\times3}+\varepsilon\Delta\boldsymbol{J} \tag{5.47}$$

对偶惯性算子的不确定性也影响重力和重力梯度力矩计算，考虑到叉乘算子的线性性，$\widehat{\boldsymbol{u}}_g^d$ 可以描述成

$$\widehat{\boldsymbol{u}}_g^d=\widehat{\overline{\boldsymbol{u}}}_g^d+\Delta\widehat{\boldsymbol{u}}_g \tag{5.48}$$

式中：

$$\widehat{\overline{\boldsymbol{u}}}_g^d=-\frac{\mu\overline{m}_d}{r_d^3}\boldsymbol{r}_d^d+\varepsilon\frac{3\mu}{r_d^5}(\boldsymbol{r}_d^d\times\overline{\boldsymbol{J}}_d\boldsymbol{r}_d^d) \tag{5.49}$$

$$\Delta\widehat{\boldsymbol{u}}_g=-\frac{\mu\Delta m}{r_d^3}\boldsymbol{r}_d^d+\varepsilon\frac{3\mu}{r_d^5}(\boldsymbol{r}_d^d\times\Delta\boldsymbol{J}\boldsymbol{r}_d^d) \tag{5.50}$$

5.2.2　基于对偶代数的姿轨一体化控制方法

分析可知，基于对偶代数建立的航天器姿轨耦合误差动力学模型是一个强耦合的非线性动力学系统，从系统和全局的角度来考虑控制问题，可依据航天器整体的运动特性，设计一体化的航天器相对轨道和姿态控制算法，并由控制算法给出期望控制力、控制力矩指令，通过控制分别的或共用的执行机构，完成航天任务。本节针对由式（5.24）和式（5.40）组成的闭环系统，设计了“前馈补偿+误差 PD”复合控制律，并进行了李雅普诺夫稳定性分析。

5.2.2.1　数学基础

为方便后续控制律设计及其稳定性分析，在 5.1.1 节的基础上进一步给出如下符号定义及相关定理。

对于对偶四元数，增加下列运算操作：

交换算子：
$$\widehat{\boldsymbol{q}}^s=\boldsymbol{q}'+\varepsilon\boldsymbol{q} \tag{5.51}$$

圆乘算子：
$$\widehat{\boldsymbol{q}}_1\circ\widehat{\boldsymbol{q}}_2=\boldsymbol{q}_1\cdot\boldsymbol{q}_2+\boldsymbol{q}_1'\cdot\boldsymbol{q}_2' \tag{5.52}$$

矩阵与对偶四元数乘法：

$$\boldsymbol{M}*\widehat{\boldsymbol{q}}=(\boldsymbol{M}_{11}\boldsymbol{q}+\boldsymbol{M}_{12}\boldsymbol{q}')+\varepsilon(\boldsymbol{M}_{21}\boldsymbol{q}+\boldsymbol{M}_{22}\boldsymbol{q}') \tag{5.53}$$

其中，

$$\boldsymbol{M}=\begin{bmatrix}\boldsymbol{M}_{11} & \boldsymbol{M}_{12}\\ \boldsymbol{M}_{21} & \boldsymbol{M}_{22}\end{bmatrix},\quad \boldsymbol{M}_{11},\boldsymbol{M}_{12},\boldsymbol{M}_{21},\boldsymbol{M}_{22}\in\mathbb{R}^{4\times4} \tag{5.54}$$

考虑到对偶矢量可看作标量部分为零的对偶四元数，因此对偶矢量可直接继承上述算子。显然，对偶矢量具有如下性质：

$$\widehat{\boldsymbol{v}}_1\times\widehat{\boldsymbol{v}}_1=\boldsymbol{0} \tag{5.55}$$

$$\widehat{\boldsymbol{v}}_1\times\widehat{\boldsymbol{v}}_2=-\widehat{\boldsymbol{v}}_2\times\widehat{\boldsymbol{v}}_1 \tag{5.56}$$

$$\hat{\boldsymbol{v}}_1^s \circ \hat{\boldsymbol{v}}_2^s = \hat{\boldsymbol{v}}_1 \circ \hat{\boldsymbol{v}}_2 \tag{5.57}$$

$$\hat{\boldsymbol{v}}_1 \circ (\hat{\boldsymbol{v}}_2 \times \hat{\boldsymbol{v}}_3) = \hat{\boldsymbol{v}}_2^s \circ (\hat{\boldsymbol{v}}_3 \times \hat{\boldsymbol{v}}_1^s) = \hat{\boldsymbol{v}}_3^s \circ (\hat{\boldsymbol{v}}_1^s \times \hat{\boldsymbol{v}}_2) \tag{5.58}$$

$$(\boldsymbol{M} * \hat{\boldsymbol{v}}_1) \circ \hat{\boldsymbol{v}}_2 = \hat{\boldsymbol{v}}_1 \circ (\boldsymbol{M}^{\mathrm{T}} * \hat{\boldsymbol{v}}_2), \quad \boldsymbol{M} \in \mathbb{R}^{6\times 6} \tag{5.59}$$

因此，利用这些符号定义可进一步将式（5.40）改写为

$$(\dot{\hat{\boldsymbol{\omega}}}_{d/t}^d)^s = \boldsymbol{M}_d^{-1} * \{ (\hat{\boldsymbol{u}}_c^d + \hat{\boldsymbol{u}}_g^d + \hat{\boldsymbol{d}}) - (\hat{\boldsymbol{\omega}}_{d/t}^d + \hat{\boldsymbol{\omega}}_{t/I}^d) \times (\boldsymbol{M}_d * ((\hat{\boldsymbol{\omega}}_{d/t}^d)^s + (\hat{\boldsymbol{\omega}}_{t/I}^d)^s)) - \boldsymbol{M}_d * (\hat{\boldsymbol{q}}_e \otimes \dot{\hat{\boldsymbol{\omega}}}_{t/I}^t \otimes \hat{\boldsymbol{q}}_e^*)^s - \boldsymbol{M}_d * (\hat{\boldsymbol{\omega}}_{t/I}^d \times \hat{\boldsymbol{\omega}}_{d/t}^d)^s \} \tag{5.60}$$

式中：对偶惯性矩阵 $\boldsymbol{M}_d$ 定义为

$$\boldsymbol{M}_d = \begin{bmatrix} m_d \boldsymbol{I}_{3\times 3} & \boldsymbol{0}_{3\times 3} \\ \boldsymbol{0}_{3\times 3} & \boldsymbol{J}_d \end{bmatrix} \tag{5.61}$$

LaSalle 不变集定理：考虑自治系统[10]

$$\dot{\boldsymbol{x}} = f(\boldsymbol{x}) \tag{5.62}$$

其中，$f{:}D\to\mathbb{R}^n$ 是从定义域 $D\subset\mathbb{R}^n$ 到 $\mathbb{R}^n$ 上的局部 Lipschitz 映射。设 $\Omega\subset D$ 是方程（5.62）的一个正不变紧集，$V{:}D\to\mathbb{R}$ 是连续可微函数，且在 Ω 内满足 $\dot{V}(\boldsymbol{x})\leqslant 0$。假设 E 是 Ω 内满足 $\dot{V}(\boldsymbol{x})=0$ 所有点的集合，$\mathcal{M}$ 是 E 内最大不变集，那么，当 $t\to\infty$ 时，始于 Ω 内的每个解都趋于 $\mathcal{M}$。

5.2.2.2 “前馈补偿+误差 PD”控制方法

分析可知，式（5.60）中的动力学模型是强耦合、非线性的，从控制简洁性、实用性以及可靠性出发，可设计如下“前馈补偿+误差 PD”控制律：

$$\hat{\boldsymbol{u}}_c^d = -k_p \mathrm{vec}((\hat{\boldsymbol{q}}_e^s - \hat{\mathbf{1}}^s) \otimes \hat{\boldsymbol{q}}_e^*) - k_d (\hat{\boldsymbol{\omega}}_{d/t}^d)^s - \hat{\boldsymbol{u}}_g^d + \boldsymbol{M}_d * (\hat{\boldsymbol{q}}_e \otimes \dot{\hat{\boldsymbol{\omega}}}_{t/I}^t \otimes \hat{\boldsymbol{q}}_e^*)^s + \hat{\boldsymbol{\omega}}_{t/I}^d \times (\boldsymbol{M}_d * (\hat{\boldsymbol{\omega}}_{t/I}^d)^s), \quad k_p, k_d > 0 \tag{5.63}$$

式中：vec(·)表示取对偶四元数的矢量部分。

对于由式（5.24）、式（5.60）和式（5.63）组成的闭环系统，系统平衡点是 $\hat{\boldsymbol{q}}_e = [0\ 0\ 0 \pm 1]^{\mathrm{T}} + \varepsilon[0\ 0\ 0\ 0]^{\mathrm{T}}$，$\hat{\boldsymbol{\omega}}_{d/t}^d = [0\ 0\ 0]^{\mathrm{T}} + \varepsilon[0\ 0\ 0]^{\mathrm{T}}$，后文中分别用 $\pm\hat{\mathbf{1}}$ 和 $\hat{\mathbf{0}}$ 表示。

针对平衡点 $\hat{\boldsymbol{q}}_e = \hat{\mathbf{1}}$，$\hat{\boldsymbol{\omega}}_{d/t}^d = \hat{\mathbf{0}}$，考虑如下 Lyapunov 函数

$$V(\hat{\boldsymbol{q}}_e, \hat{\boldsymbol{\omega}}_{d/t}^d) = k_p (\hat{\boldsymbol{q}}_e - \hat{\mathbf{1}}) \circ (\hat{\boldsymbol{q}}_e - \hat{\mathbf{1}}) + \frac{1}{2} (\hat{\boldsymbol{\omega}}_{d/t}^d)^s \circ (\boldsymbol{M}_d * (\hat{\boldsymbol{\omega}}_{d/t}^d)^s) \tag{5.64}$$

显然，$V(\hat{\boldsymbol{q}}_e = \hat{\mathbf{1}}, \hat{\boldsymbol{\omega}}_{d/t}^d = \hat{\mathbf{0}}) = 0$，且对任何非平衡点均有 $V(\hat{\boldsymbol{q}}_e, \hat{\boldsymbol{\omega}}_{d/t}^d) > 0$。

对 V 进行求导可得

$$\dot{V}(\hat{\boldsymbol{q}}_e,\hat{\boldsymbol{\omega}}_{d/t}^d)=2k_p(\hat{\boldsymbol{q}}_e-\hat{\mathbf{1}})\circ\dot{\hat{\boldsymbol{q}}}_e+(\hat{\boldsymbol{\omega}}_{d/t}^d)^s\circ(\boldsymbol{M}_d*(\dot{\hat{\boldsymbol{\omega}}}_{d/t}^d)^s) \tag{5.65}$$

将式（5.24）和式（5.60）代入式（5.65），并利用式（5.58）可得

$$\begin{aligned}\dot{V}=(\hat{\boldsymbol{\omega}}_{d/t}^d)^s\circ\{k_p(\hat{\boldsymbol{q}}_e^s-\hat{\mathbf{1}}^s)\otimes\hat{\boldsymbol{q}}_e^*+\hat{\boldsymbol{u}}_c^d+\hat{\boldsymbol{u}}_g^d-(\hat{\boldsymbol{\omega}}_{d/t}^d+\hat{\boldsymbol{\omega}}_{t/I}^d)\times(\boldsymbol{M}_d*((\hat{\boldsymbol{\omega}}_{d/t}^d)^s+(\hat{\boldsymbol{\omega}}_{t/I}^d)^s))-\\ \boldsymbol{M}_d*(\hat{\boldsymbol{q}}_e\otimes\dot{\hat{\boldsymbol{\omega}}}_{t/I}^t\otimes\hat{\boldsymbol{q}}_e^*)^s-\boldsymbol{M}_d*(\hat{\boldsymbol{\omega}}_{t/I}^d\times\hat{\boldsymbol{\omega}}_{d/t}^d)^s\}\end{aligned} \tag{5.66}$$

进一步，将式（5.63）中的控制律代入式（5.66），可得

$$\begin{aligned}\dot{V}=&(\hat{\boldsymbol{\omega}}_{d/t}^d)^s\circ(-k_d(\hat{\boldsymbol{\omega}}_{d/t}^d)^s)+(\hat{\boldsymbol{\omega}}_{d/t}^d)^s\circ(k_p(\hat{\boldsymbol{q}}_e^s-\hat{\mathbf{1}}^s)\otimes\hat{\boldsymbol{q}}_e^*-\\ &k_p\mathrm{vec}((\hat{\boldsymbol{q}}_e^s-\hat{\mathbf{1}}^s)\otimes\hat{\boldsymbol{q}}_e^*))+(\hat{\boldsymbol{\omega}}_{d/t}^d)^s\circ\{-(\hat{\boldsymbol{\omega}}_{d/t}^d+\hat{\boldsymbol{\omega}}_{t/I}^d)\times\\ &(\boldsymbol{M}_d*((\hat{\boldsymbol{\omega}}_{d/t}^d)^s+(\hat{\boldsymbol{\omega}}_{t/I}^d)^s))-\boldsymbol{M}_d*(\hat{\boldsymbol{\omega}}_{t/I}^d\times\hat{\boldsymbol{\omega}}_{d/t}^d)^s+\\ &\hat{\boldsymbol{\omega}}_{t/I}^d\times(\boldsymbol{M}_d*(\hat{\boldsymbol{\omega}}_{t/I}^d)^s)\}\end{aligned} \tag{5.67}$$

分析可知，式（5.67）中的第二项为零，且第三项可以改写为

$$\begin{aligned}&(\hat{\boldsymbol{\omega}}_{d/t}^d)^s\circ\{-(\hat{\boldsymbol{\omega}}_{d/t}^d+\hat{\boldsymbol{\omega}}_{t/I}^d)\times(\boldsymbol{M}_d*((\hat{\boldsymbol{\omega}}_{d/t}^d)^s+(\hat{\boldsymbol{\omega}}_{t/I}^d)^s))-\\ &\boldsymbol{M}_d*(\hat{\boldsymbol{\omega}}_{t/I}^d\times\hat{\boldsymbol{\omega}}_{d/t}^d)^s+\hat{\boldsymbol{\omega}}_{t/I}^d\times(\boldsymbol{M}_d*(\hat{\boldsymbol{\omega}}_{t/I}^d)^s)\}\\ =&((\hat{\boldsymbol{\omega}}_{d/I}^d)^s-(\hat{\boldsymbol{\omega}}_{t/I}^d)^s)\circ\{-\hat{\boldsymbol{\omega}}_{d/I}^d\times(\boldsymbol{M}_d*(\hat{\boldsymbol{\omega}}_{d/I}^d)^s)-\\ &\boldsymbol{M}_d*(\hat{\boldsymbol{\omega}}_{t/I}^d\times(\hat{\boldsymbol{\omega}}_{d/I}^d-\hat{\boldsymbol{\omega}}_{t/I}^d))^s+\hat{\boldsymbol{\omega}}_{t/I}^d\times(\boldsymbol{M}_d*(\hat{\boldsymbol{\omega}}_{t/I}^d)^s)\}\\ =&(\hat{\boldsymbol{\omega}}_{d/I}^d)^s\circ\{-\hat{\boldsymbol{\omega}}_{d/I}^d\times(\boldsymbol{M}_d*(\hat{\boldsymbol{\omega}}_{d/I}^d)^s)-\boldsymbol{M}_d*(\hat{\boldsymbol{\omega}}_{t/I}^d\times\hat{\boldsymbol{\omega}}_{d/I}^d)^s+\hat{\boldsymbol{\omega}}_{t/I}^d\times(\boldsymbol{M}_d*(\hat{\boldsymbol{\omega}}_{t/I}^d)^s)\}-\\ &(\hat{\boldsymbol{\omega}}_{t/I}^d)^s\circ\{-\hat{\boldsymbol{\omega}}_{d/I}^d\times(\boldsymbol{M}_d*(\hat{\boldsymbol{\omega}}_{d/I}^d)^s)-\boldsymbol{M}_d*(\hat{\boldsymbol{\omega}}_{t/I}^d\times\hat{\boldsymbol{\omega}}_{d/I}^d)^s+\hat{\boldsymbol{\omega}}_{t/I}^d\times(\boldsymbol{M}_d*(\hat{\boldsymbol{\omega}}_{t/I}^d)^s)\}\\ =&-(\hat{\boldsymbol{\omega}}_{d/I}^d)^s\circ(\hat{\boldsymbol{\omega}}_{d/I}^d\times(\boldsymbol{M}_d*(\hat{\boldsymbol{\omega}}_{d/I}^d)^s))-(\hat{\boldsymbol{\omega}}_{d/I}^d)^s\circ(\boldsymbol{M}_d*(\hat{\boldsymbol{\omega}}_{t/I}^d\times\hat{\boldsymbol{\omega}}_{d/I}^d)^s)+\\ &(\hat{\boldsymbol{\omega}}_{d/I}^d)^s\circ(\hat{\boldsymbol{\omega}}_{t/I}^d\times(\boldsymbol{M}_d*(\hat{\boldsymbol{\omega}}_{t/I}^d)^s))+(\hat{\boldsymbol{\omega}}_{t/I}^d)^s\circ(\hat{\boldsymbol{\omega}}_{d/I}^d\times(\boldsymbol{M}_d*(\hat{\boldsymbol{\omega}}_{d/I}^d)^s))+\\ &(\hat{\boldsymbol{\omega}}_{t/I}^d)^s\circ(\boldsymbol{M}_d*(\hat{\boldsymbol{\omega}}_{t/I}^d\times\hat{\boldsymbol{\omega}}_{d/I}^d)^s)-(\hat{\boldsymbol{\omega}}_{t/I}^d)^s\circ(\hat{\boldsymbol{\omega}}_{t/I}^d\times(\boldsymbol{M}_d*(\hat{\boldsymbol{\omega}}_{t/I}^d)^s))\end{aligned} \tag{5.68}$$

根据式（5.55）和式（5.58），可知式（5.68）中的第一项和最后一项为零。进一步，利用式（5.57）和式（5.59），可将式（5.68）中的第二项至第五项改写为

$$\begin{aligned}&-(\boldsymbol{M}_d*(\hat{\boldsymbol{\omega}}_{d/I}^d)^s)^s\circ(\hat{\boldsymbol{\omega}}_{t/I}^d\times\hat{\boldsymbol{\omega}}_{d/I}^d)+(\hat{\boldsymbol{\omega}}_{d/I}^d)^s\circ(\hat{\boldsymbol{\omega}}_{t/I}^d\times(\boldsymbol{M}_d*(\hat{\boldsymbol{\omega}}_{t/I}^d)^s))+\\ &(\hat{\boldsymbol{\omega}}_{t/I}^d)^s\circ(\hat{\boldsymbol{\omega}}_{d/I}^d\times(\boldsymbol{M}_d*(\hat{\boldsymbol{\omega}}_{d/I}^d)^s))+(\boldsymbol{M}_d*(\hat{\boldsymbol{\omega}}_{t/I}^d)^s)^s\circ(\hat{\boldsymbol{\omega}}_{t/I}^d\times\hat{\boldsymbol{\omega}}_{d/I}^d)\end{aligned} \tag{5.69}$$

对于式（5.69）中的第一项和最后一项，利用式（5.56）和式（5.58），可进一步将式（5.69）改写为

$$-(\hat{\boldsymbol{\omega}}_{t/I}^{d})^{s}\circ(\hat{\boldsymbol{\omega}}_{d/I}^{d}\times(\boldsymbol{M}_{d}*(\hat{\boldsymbol{\omega}}_{d/I}^{d})^{s}))+(\hat{\boldsymbol{\omega}}_{d/I}^{d})^{s}\circ(\hat{\boldsymbol{\omega}}_{t/I}^{d}\times(\boldsymbol{M}_{d}*(\hat{\boldsymbol{\omega}}_{t/I}^{d})^{s}))+$$
$$(\hat{\boldsymbol{\omega}}_{t/I}^{d})^{s}\circ(\hat{\boldsymbol{\omega}}_{d/I}^{d}\times(\boldsymbol{M}_{d}*(\hat{\boldsymbol{\omega}}_{d/I}^{d})^{s}))-(\hat{\boldsymbol{\omega}}_{d/I}^{d})^{s}\circ(\hat{\boldsymbol{\omega}}_{t/I}^{d}\times(\boldsymbol{M}_{d}*(\hat{\boldsymbol{\omega}}_{t/I}^{d})^{s}))=0 \tag{5.70}$$

于是，有

$$\dot{V}=-k_{d}(\hat{\boldsymbol{\omega}}_{d/t}^{d})^{s}\circ(\hat{\boldsymbol{\omega}}_{d/t}^{d})^{s}\leqslant 0 \tag{5.71}$$

这表明$\dot{V}$是负半定的。由 LaSalle 不变集定理可知，所有的点收敛到闭环系统的最大的不变集 $\mathcal{M}=\{(\hat{\boldsymbol{q}}_{e},\hat{\boldsymbol{\omega}}_{d/t}^{d})\,|\,\dot{V}=0\}=\{(\hat{\boldsymbol{q}}_{e},\hat{\boldsymbol{\omega}}_{d/t}^{d})\,|\,\hat{\boldsymbol{\omega}}_{d/t}^{d}=\hat{\boldsymbol{0}}\}$。根据式（5.60）和式（5.63）可知，在这个最大的不变集内，有 $\mathrm{vec}((\hat{\boldsymbol{q}}_{e}^{s}-\hat{\boldsymbol{1}}^{s})\otimes\hat{\boldsymbol{q}}_{e}^{*})=\hat{\boldsymbol{0}}$。对于表达式$(\hat{\boldsymbol{q}}_{e}^{s}-\hat{\boldsymbol{1}}^{s})\otimes\hat{\boldsymbol{q}}_{e}^{*}$，计算可得

$$\begin{aligned}(\hat{\boldsymbol{q}}_{e}^{s}-\hat{\boldsymbol{1}}^{s})\otimes\hat{\boldsymbol{q}}_{e}^{*}&=\left(\frac{1}{2}\boldsymbol{p}_{d/t}^{d}\otimes\boldsymbol{q}_{e}+\varepsilon(\boldsymbol{q}_{e}-\boldsymbol{1})\right)\otimes\left(\boldsymbol{q}_{e}^{*}-\varepsilon\frac{1}{2}\boldsymbol{q}_{e}^{*}\otimes\boldsymbol{p}_{d/t}^{d}\right)\\&=\frac{1}{2}\boldsymbol{p}_{d/t}^{d}+\varepsilon\left(\boldsymbol{1}-\boldsymbol{q}_{e}^{*}-\frac{1}{4}\|\boldsymbol{p}_{d/t}^{d}\|^{2}\right)\end{aligned} \tag{5.72}$$

对式（5.72）取矢部，可得

$$\begin{aligned}\mathrm{vec}((\hat{\boldsymbol{q}}_{e}^{s}-\hat{\boldsymbol{1}}^{s})\otimes\hat{\boldsymbol{q}}_{e}^{*})&=\mathrm{vec}\left(\frac{1}{2}\boldsymbol{p}_{d/t}^{d}\right)\varepsilon\left(\boldsymbol{1}-\boldsymbol{q}_{e}^{*}-\frac{1}{4}\|\boldsymbol{p}_{d/t}^{d}\|^{2}\right)\\&=\frac{1}{2}\boldsymbol{p}_{d/t}^{d}+\varepsilon\,\mathrm{vec}(-\boldsymbol{q}_{e}^{*})\\&=\frac{1}{2}\boldsymbol{p}_{d/t}^{d}+\varepsilon\,\mathrm{vec}(\boldsymbol{q}_{e})\end{aligned} \tag{5.73}$$

因此，要满足 $\mathrm{vec}((\hat{\boldsymbol{q}}_{e}^{s}-\hat{\boldsymbol{1}}^{s})\otimes\hat{\boldsymbol{q}}_{e}^{*})=\hat{\boldsymbol{0}}$，当且仅当$\frac{1}{2}\boldsymbol{p}_{d/t}^{d}=\boldsymbol{0}$ 和$\boldsymbol{Q}_{e}=\boldsymbol{0}$。对于后面的条件，有 $\boldsymbol{q}_{e}=\pm\boldsymbol{1}$，进一步结合$\frac{1}{2}\boldsymbol{p}_{d/t}^{d}=\boldsymbol{0}$，可得$\hat{\boldsymbol{q}}_{e}=\pm\hat{\boldsymbol{1}}$。因此，闭环系统最大的不变集为 $\mathcal{M}=\{(\hat{\boldsymbol{q}}_{e},\hat{\boldsymbol{\omega}}_{d/t}^{d})\,|\,\dot{V}=0\}=\{(\hat{\boldsymbol{q}}_{e},\hat{\boldsymbol{\omega}}_{d/t}^{d})\,|\,\hat{\boldsymbol{q}}_{e}=\pm\hat{\boldsymbol{1}},\hat{\boldsymbol{\omega}}_{d/t}^{d}=\hat{\boldsymbol{0}}\}$，即平衡点。由于所选定的 Lyapunov 函数是径向无界的，所以闭环系统是全局渐近稳定的，证毕。

对于式（5.63）中的控制增益系数 k_p 和 k_d，可结合经典控制理论考虑如下矩阵形式：

$$\boldsymbol{K}_{p}=\begin{bmatrix}m_{d}\omega_{r}^{2}\boldsymbol{I}_{3\times3} & \boldsymbol{0}_{3\times3}\\ \boldsymbol{0}_{3\times3} & \omega_{\theta}^{2}\boldsymbol{J}_{d}\end{bmatrix},\quad \boldsymbol{K}_{d}=\begin{bmatrix}2m_{d}\zeta_{r}\omega_{r}\boldsymbol{I}_{3\times3} & \boldsymbol{0}_{3\times3}\\ \boldsymbol{0}_{3\times3} & 2\zeta_{\theta}\omega_{\theta}\boldsymbol{J}_{d}\end{bmatrix} \tag{5.74}$$

式中：ω_r 和 ω_θ 为自然频率；ζ_r 和 ζ_θ 为阻尼比。

因此，式（5.63）中的控制律可以改写为

$$\hat{\boldsymbol{u}}_c^d=-\boldsymbol{K}_p*\mathrm{vec}((\hat{\boldsymbol{q}}_e^s-\hat{\boldsymbol{1}}^s)\otimes\hat{\boldsymbol{q}}_e^*)-\boldsymbol{K}_d*(\hat{\boldsymbol{\omega}}_{d/t}^d)^s-\hat{\boldsymbol{u}}_g^d+\boldsymbol{M}_d*(\hat{\boldsymbol{q}}_e\otimes\dot{\hat{\boldsymbol{\omega}}}_{t/I}^t\otimes\hat{\boldsymbol{q}}_e^*)^s+\hat{\boldsymbol{\omega}}_{t/I}^d\times(\boldsymbol{M}_d*(\hat{\boldsymbol{\omega}}_{t/I}^d)^s) \tag{5.75}$$

可以看出，利用 Lyapunov 定理证明上述相对位置和姿态耦合控制器只需一步，并不需要将其拆分为位置控制和姿态控制两个子问题，且选取的 Lyapunov 函数形式与传统的姿态控制器证明选取的类似。

5.2.3　全推力器控制系统设计

5.2.3.1　推力器构型设计

推力器构型设计是在系统控制要求与约束条件范围内，合理地选择推力器的数目、类型，并且确定各推力器在航天器上的安装方位。根据任务需要，从星上的执行机构采用 12 个小推力器用于轨道控制和姿态控制，实现全推力器姿轨一体化控制，推力器安装配置如图 5.2 所示。

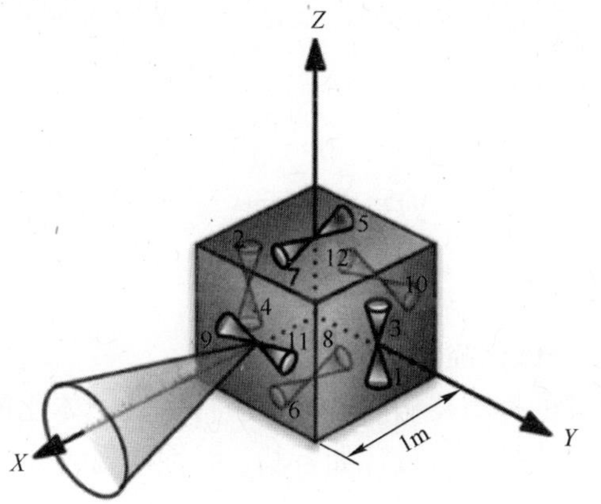

图 5.2　从星推力器配置

注意到推力器系统都是成对安装的，可用来同时提供力和力矩。推力器的单位推力方向和安装位置矢量可参见矩阵 $\boldsymbol{A}$ 和 $\boldsymbol{R}$。

$$\boldsymbol{A}=\begin{bmatrix}0&0&0&0&1&-1&-1&1&0&0&0&0\\0&0&0&0&0&0&0&0&1&-1&-1&1\\1&-1&-1&1&0&0&0&0&0&0&0&0\end{bmatrix} \tag{5.76}$$

$$\boldsymbol{R}=\begin{bmatrix}0&0&0&0&0&0&0&0&0.5&-0.5&0.5&-0.5\\0.5&-0.5&0.5&-0.5&0&0&0&0&0&0&0&0\\0&0&0&0&0.5&-0.5&0.5&-0.5&0&0&0&0\end{bmatrix}\tag{5.77}$$

其中，矩阵 $\boldsymbol{A}$ 和 $\boldsymbol{R}$ 中的第 i 列对应于第 i 个推力器的单位推力矢量方向与安装位置。因此，第 i 个推力器产生的推力和力矩可分别表示为

$$\boldsymbol{f}_i=u_i\boldsymbol{a}_i\tag{5.78}$$

$$\boldsymbol{\tau}_i=(\boldsymbol{r}_i\times\boldsymbol{a}_i)u_i\tag{5.79}$$

式中：u_i 为第 i 个推力器的推力大小；$\boldsymbol{a}_i$ 和 $\boldsymbol{r}_i$ 分别对应于矩阵 $\boldsymbol{A}$ 和 $\boldsymbol{R}$ 的第 i 列。定义矢量 $\boldsymbol{u}=[u_1,u_2,\cdots,u_{12}]^{\mathrm{T}}$ 包含所有推力器推力大小，则由所有推力器产生的力和力矩总大小分别为

$$\boldsymbol{f}=\sum_{i=1}^{12}\boldsymbol{f}_i=\sum_{i=1}^{12}u_i\boldsymbol{a}_i=\boldsymbol{A}\boldsymbol{u}\tag{5.80}$$

$$\boldsymbol{\tau}=\sum_{i=1}^{12}\boldsymbol{\tau}_i=\sum_{i=1}^{12}(\boldsymbol{r}_i\times\boldsymbol{a}_i)u_i=\boldsymbol{B}\boldsymbol{u}\tag{5.81}$$

式中：

$$\begin{aligned}\boldsymbol{B}&=[\boldsymbol{r}_1\times\boldsymbol{a}_1,\boldsymbol{r}_2\times\boldsymbol{a}_2,\cdots,\boldsymbol{r}_{12}\times\boldsymbol{a}_{12}]\\&=\begin{bmatrix}0.5&0.5&-0.5&-0.5&0&0&0&0&0&0&0&0\\0&0&0&0&0.5&0.5&-0.5&-0.5&0&0&0&0\\0&0&0&0&0&0&0&0&0.5&0.5&-0.5&-0.5\end{bmatrix}\end{aligned}\tag{5.82}$$

5.2.3.2 姿轨一体化控制分配算法设计

结合前一节设计的推力器构型，本节进行姿轨一体化控制分配算法设计。以工质消耗最少为最优指标，设计如下基于线性规划的分配算法模型：

$$\begin{gathered}\min_{u} J=\boldsymbol{e}^{\mathrm{T}}\boldsymbol{u}\\ \text{s. t.}\begin{cases}\begin{bmatrix}\boldsymbol{f}_c\\\boldsymbol{\tau}_c\end{bmatrix}=\begin{bmatrix}\boldsymbol{A}\\\boldsymbol{B}\end{bmatrix}\boldsymbol{u}\equiv\boldsymbol{D}\boldsymbol{u}\\\boldsymbol{u}_{\min}\leqslant\boldsymbol{u}\leqslant\boldsymbol{u}_{\max}\end{cases}\end{gathered}\tag{5.83}$$

式中：$\boldsymbol{e}=[1,\cdots,1]^{\mathrm{T}}\in\mathbb{R}^{12\times1}$ 为指标权重系数；$\boldsymbol{u}$ 为待求解的控制指令列矢量；$\boldsymbol{f}_c$ 和 $\boldsymbol{\tau}_c$ 分别为控制律设计得到的期望控制力和力矩矢量；矩阵 $\boldsymbol{A}$ 和 $\boldsymbol{B}$ 分别为单位推力和力矩矩阵，由前一节设计所得；$\boldsymbol{u}_{\min}\in\mathbb{R}^{12\times1}$ 和 $\boldsymbol{u}_{\max}\in\mathbb{R}^{12\times1}$ 分别为控制变量下限和上限。对于该模型，在 MATLAB 中可直接运行函数 linprog 求解最优解。

5.3 仿真实验与分析

本节以空间交会对接任务为例，验证所设计的基于对偶代数的姿轨一体化控制方法以及全推力器控制分配算法。基本仿真参数设置如下：主星质量 $m_c=20\text{kg}$，惯量张量 $\boldsymbol{J}_c=\text{diag}[8,5,4]\ \text{kg}\cdot\text{m}^2$；主星轨道六要素：半长轴 $a=6998455\text{m}$，偏心率 $e=0.1$，轨道倾角 $i=30°$，升交点角距 $\Omega=0$，近地点角距 $\omega=0$，真近点角 $\theta=0$。从星的标称质量 $\overline{m}_d=20\text{kg}$，质量不确定性干扰 $\Delta m=2\text{kg}$，标称惯量张量阵和惯量张量不确定性分别为

$$\overline{\boldsymbol{J}}_d=\begin{bmatrix}8&0&0\\0&5&0\\0&0&4\end{bmatrix}\text{kg}\cdot\text{m}^2,\quad \Delta\boldsymbol{J}=\begin{bmatrix}1.6&0&0\\0&1.0&0\\0&0&0.8\end{bmatrix}\text{kg}\cdot\text{m}^2 \tag{5.84}$$

从星上的执行机构采用 5.2.3 节中的全推力器构型设计，并利用 PWPF 技术将开关型控制量调制成连续性控制量，推力幅值满足 $u_{\min}=0$，$u_{\max}=0.1\text{N}$。

作用在从星上的外部对偶干扰力矢量为

$$\begin{aligned}\hat{\boldsymbol{d}}=&[0.001\quad 0.001\quad 0.001]^{\text{T}}\sin(t)\text{N}+\\&\varepsilon[0.005\quad 0.005\quad 0.005]^{\text{T}}\sin(t)\text{N}\cdot\text{m}\end{aligned} \tag{5.85}$$

假设初始时刻从星的惯性角速度为 $\boldsymbol{\omega}_{d/I}^{d}(t_0)=[-0.002,0,0.0011]^{\text{T}}\text{rad/s}$，相对于主星 LVLH 系的姿态四元数为 $\boldsymbol{q}_{d/H}(t_0)=[0.4\quad 0.2\quad -0.4\quad 0.8]^{\text{T}}$；初始相对位置和相对速度矢量分别为 $\boldsymbol{p}_{d/c}^{H}(t_0)=[-10\quad -30\quad 0]^{\text{T}}\text{m}$ 和 $\dot{\boldsymbol{p}}_{d/c}^{H}(t_0)=[0\quad 0\quad 0]^{\text{T}}\text{m/s}$。这些参数均可由第 4 章中的相对导航方法实时地在线计算得到。

5.3.1 仿真算例 1

本节假设对主星施加控制使其体系与 LVLH 系时刻重合，则选择主星的 H 系为期望坐标系 t 系，基于对偶代数的相对运动学和动力学方程分别为

$$\dot{\hat{\boldsymbol{q}}}_{d/H}=\frac{1}{2}\hat{\boldsymbol{\omega}}_{d/H}^{d}\otimes\hat{\boldsymbol{q}}_{d/H} \tag{5.86}$$

$$\begin{aligned}(\dot{\hat{\boldsymbol{\omega}}}_{d/H}^{d})^{s}=\boldsymbol{M}_d^{-1}*\{&(\hat{\boldsymbol{u}}_c^{d}+\hat{\boldsymbol{u}}_g^{d}+\hat{\boldsymbol{d}})-(\hat{\boldsymbol{\omega}}_{d/H}^{d}+\hat{\boldsymbol{\omega}}_{H/I}^{d})\times(\boldsymbol{M}_d*((\hat{\boldsymbol{\omega}}_{d/H}^{d})^{s}+(\hat{\boldsymbol{\omega}}_{H/I}^{d})^{s}))-\\&\boldsymbol{M}_d*(\hat{\boldsymbol{q}}_{d/H}\otimes\dot{\hat{\boldsymbol{\omega}}}_{H/I}^{H}\otimes\hat{\boldsymbol{q}}_{d/H}^{*})^{s}-\boldsymbol{M}_d*(\hat{\boldsymbol{\omega}}_{H/I}^{d}\times\hat{\boldsymbol{\omega}}_{d/H}^{d})^{s}\}\end{aligned} \tag{5.87}$$

式中：变量$\hat{\boldsymbol{\omega}}_{H/I}^{H}$、$\hat{\boldsymbol{\omega}}_{H/I}^{d}$和$\dot{\hat{\boldsymbol{\omega}}}_{H/I}^{H}$计算公式为

$$\hat{\boldsymbol{\omega}}_{H/I}^{H}=\boldsymbol{\omega}_{H/I}^{H}+\varepsilon(\dot{\boldsymbol{r}}_{H/I}^{H}+\boldsymbol{\omega}_{H/I}^{H}\times\boldsymbol{r}_{H/I}^{H}) \tag{5.88}$$

$$\hat{\boldsymbol{\omega}}_{H/I}^{d}=\hat{\boldsymbol{q}}_{d/H}\otimes\hat{\boldsymbol{\omega}}_{H/I}^{H}\otimes\hat{\boldsymbol{q}}_{d/H}^{*} \tag{5.89}$$

$$\dot{\hat{\boldsymbol{\omega}}}_{H/I}^{H}=\dot{\boldsymbol{\omega}}_{H/I}^{H}+\varepsilon(\ddot{\boldsymbol{r}}_{H/I}^{H}+\dot{\boldsymbol{\omega}}_{H/I}^{H}\times\boldsymbol{r}_{H/I}^{H}+\boldsymbol{\omega}_{H/I}^{H}\times\dot{\boldsymbol{r}}_{H/I}^{H}) \tag{5.90}$$

由坐标系定义可知，$\boldsymbol{r}_{H/I}^{H}=[r_c \quad 0 \quad 0]^{\mathrm{T}}$，$\dot{\boldsymbol{r}}_{H/I}^{H}=[\dot{r}_c \quad 0 \quad 0]^{\mathrm{T}}$，$\ddot{\boldsymbol{r}}_{H/I}^{H}=[\ddot{r}_c \quad 0 \quad 0]^{\mathrm{T}}$，$\boldsymbol{\omega}_{H/I}^{H}=[0 \quad 0 \quad \dot{\theta}]^{\mathrm{T}}$，$\dot{\boldsymbol{\omega}}_{H/I}^{H}=[0 \quad 0 \quad \ddot{\theta}]^{\mathrm{T}}$。其中，$r_c$和$\dot{\theta}$分别为主星的地心距和真近点角速度大小，这些轨道元素可通过式（2.3）积分或者第4章中的相对导航方法计算得到。由主星的轨道参数设置可计算这些变量的初始值分别为$r_c(t_0)=a(1-e)$，$\dot{r}_c(t_0)=0$，$\theta(t_0)=0$，$\dot{\theta}(t_0)=\sqrt{\mu/p}(1+e)/r_c$。

由于主星的LVLH系固连在主星质心上，所以根据前述设置的从星初始参数可知$\boldsymbol{p}_{d/H}^{H}=[-10 \quad -30 \quad 0]^{\mathrm{T}}\mathrm{m}$和$\dot{\boldsymbol{p}}_{d/H}^{H}=[0 \quad 0 \quad 0]^{\mathrm{T}}\mathrm{m/s}$。那么初始的对偶四元数$\hat{\boldsymbol{q}}_{d/H}$以及旋量$\hat{\boldsymbol{\omega}}_{d/H}^{d}$计算公式为

$$\hat{\boldsymbol{q}}_{d/H}=\boldsymbol{q}_{d/H}+\varepsilon\,\frac{1}{2}\boldsymbol{q}_{d/H}\otimes\boldsymbol{p}_{d/H}^{H} \tag{5.91}$$

$$\hat{\boldsymbol{\omega}}_{d/H}^{d}=\boldsymbol{\omega}_{d/H}^{d}+\varepsilon(\boldsymbol{A}_{H}^{d}(\boldsymbol{q}_{d/H})\dot{\boldsymbol{p}}_{d/H}^{H}) \tag{5.92}$$

式中：$\boldsymbol{\omega}_{d/H}^{d}=\boldsymbol{\omega}_{d/I}^{d}-\boldsymbol{A}_{H}^{d}(\boldsymbol{q}_{d/H})\boldsymbol{\omega}_{H/I}^{H}$。事实上，结合第4章中的相对导航方法以及式（5.88）~式（5.92），可实时地为式（5.63）或式（5.75）提供所需参数值，与利用式（5.86）和式（5.87）积分计算$\hat{\boldsymbol{q}}_{d/H}$和$\hat{\boldsymbol{\omega}}_{d/H}^{d}$相比，该方法提供的实时导航值更为精确，且可进行导航控制一体化仿真。

对于式（5.63）或式（5.75）中$\hat{\boldsymbol{u}}_{g}^{d}$估值的计算，可通过先求解$\hat{\boldsymbol{q}}_{H/I}$，然后由式$\hat{\boldsymbol{q}}_{d/I}=\hat{\boldsymbol{q}}_{d/H}\otimes\hat{\boldsymbol{q}}_{H/I}$计算$\hat{\boldsymbol{q}}_{d/I}$，进一步结合标称对偶惯性矩阵$\overline{\boldsymbol{M}}_{d}$计算得到$\hat{\boldsymbol{u}}_{g}^{d}$的估计值。至于$\hat{\boldsymbol{q}}_{H/I}$，由坐标系转换知识可知，从$H$系到$I$系的姿态转换阵$\boldsymbol{A}_{H}^{I}$为

$$\boldsymbol{A}_{H}^{I}=\boldsymbol{R}_{3}(-\Omega)\boldsymbol{R}_{1}(-i)\boldsymbol{R}_{3}(-(\omega+\theta)) \tag{5.93}$$

式中：$\boldsymbol{R}_{1}(\cdot)$和$\boldsymbol{R}_{3}(\cdot)$分别是绕$x$轴和$z$轴的初等姿态转换矩阵。因此，可由式（5.93）计算得到$\boldsymbol{A}_{I}^{H}$，并进一步反解得到四元数$\boldsymbol{q}_{H/I}$，于是对偶四元数$\hat{\boldsymbol{q}}_{H/I}$可由公式$\hat{\boldsymbol{q}}_{H/I}=\boldsymbol{q}_{H/I}+\varepsilon\,\frac{1}{2}\boldsymbol{r}_{H/I}^{H}\otimes\boldsymbol{q}_{H/I}$计算得到。

仿真参数设置如前所述，采用式（5.75）中的姿轨一体化控制律，控制参数选取为$\omega_r=\omega_\theta=1/30$，$\zeta_r=\zeta_\theta=1.0$，并利用5.2.3节中的控制分配算法实现全推力器姿轨一体化控制，仿真时间为300s。仿真结果如图5.3~图5.9所示。

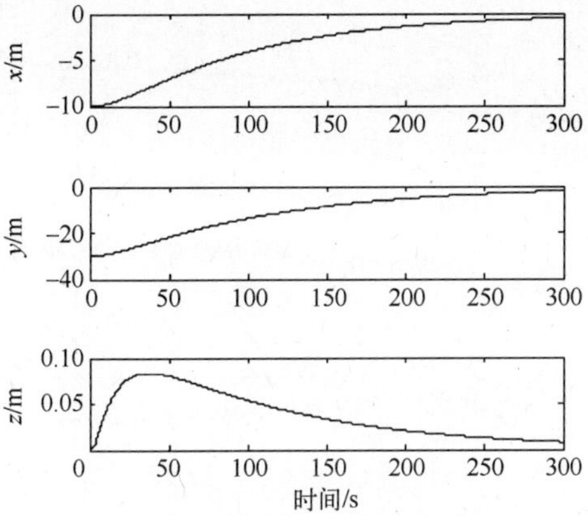

图 5.3　相对位置 $\boldsymbol{p}_{d/H}^{H}$

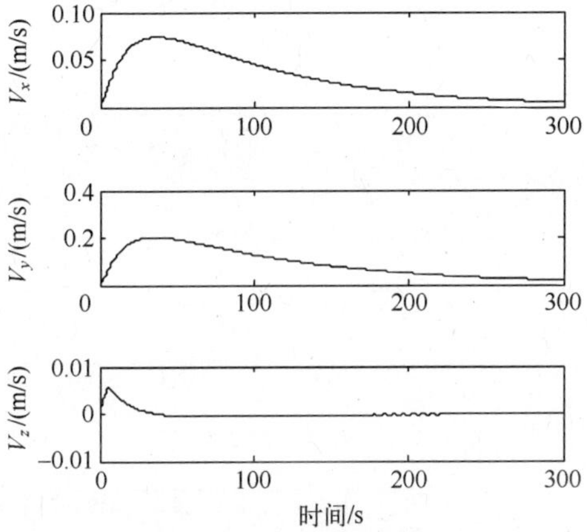

图 5.4　相对速度 $\dot{\boldsymbol{p}}_{d/H}^{H}$

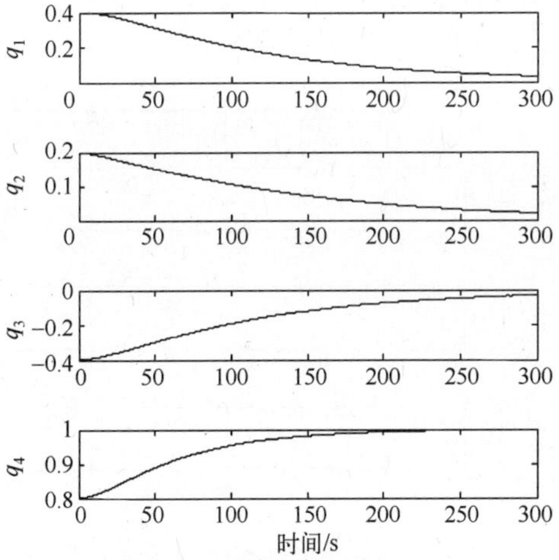

图 5.5　相对姿态四元数 $\boldsymbol{q}_{d/H}$

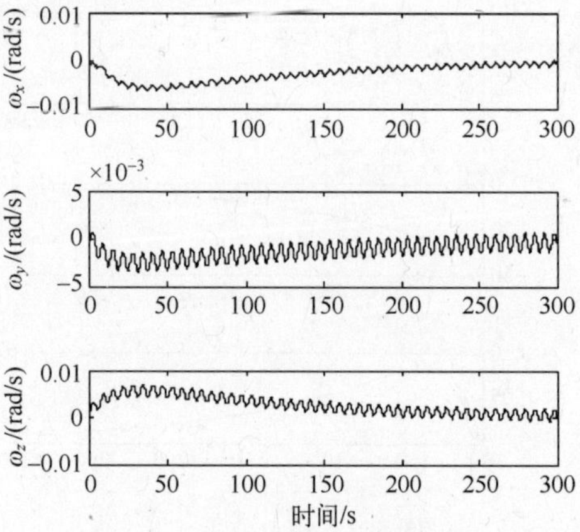

图 5.6　相对姿态角速度 $\boldsymbol{\omega}_{d/H}^{d}$

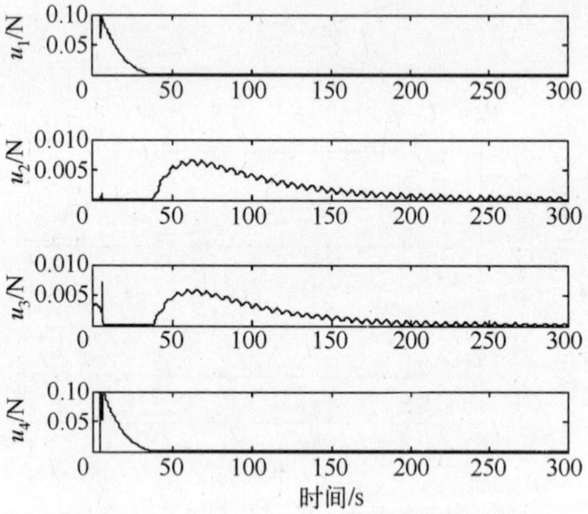

图 5.7　1~4 号推力器指令变化

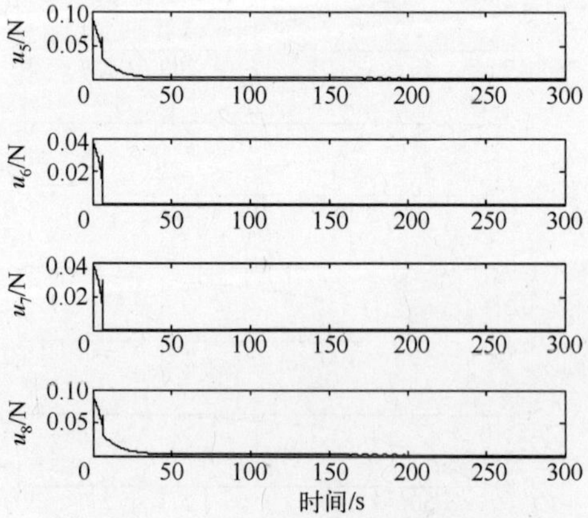

图 5.8　5~8 号推力器指令变化

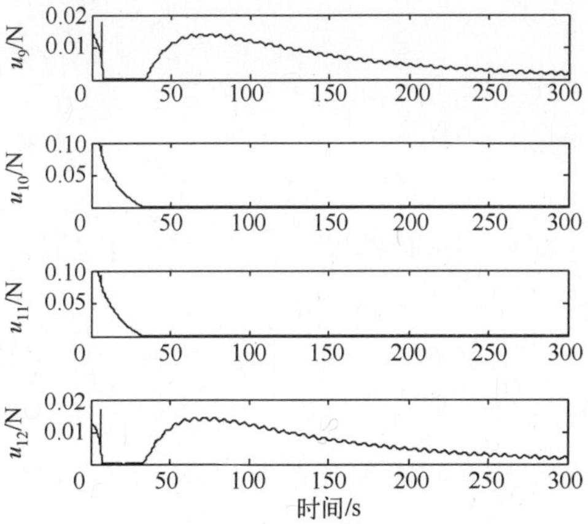

图 5.9　9~12 号推力器指令变化

图 5.3 和图 5.4 分别为相对位置和相对速度曲线，它们均描述在主星 LVLH 系中。可以看出，相对位置和速度指向变化平稳，在 $t=300\text{s}$ 时，三轴的相对位置精度控制在 0.1m 左右，相对速度控制在 $5\times10^{-3}\text{m/s}$ 以内。图 5.5 和图 5.6 分别给出相对姿态四元数 $\boldsymbol{q}_{d/H}$ 和相对姿态角速度 $\boldsymbol{\omega}_{d/H}^{d}$ 的变化曲线。虽然姿态角速度一直在振荡，但其趋势是收敛的，且最后相对姿态角速度控制在 $5\times10^{-4}\text{rad/s}$ 左右，相对姿态控制在 0.01°左右，满足空间自主逼近的控制精度和稳定度要求。这表明设计的姿轨一体化控制算法是有效的，并且对质量和转动惯量的不确定性以及外界扰动具有较好的鲁棒性。图 5.7~图 5.9 分别给出了控制过程中 1~12 号推力器的控制指令序列，各推力器控制输出均控制在 0.1N 以内。

5.3.2　仿真算例 2

本节假设对主星不施加任何控制，且其体系与 LVLH 系不重合，则选择主星的体系 c 系为期望坐标系 t 系，那么基于对偶代数的相对运动学和动力学方程分别为

$$\dot{\hat{\boldsymbol{q}}}_{d/c}=\frac{1}{2}\hat{\boldsymbol{\omega}}_{d/c}^{d}\otimes\hat{\boldsymbol{q}}_{d/c} \tag{5.94}$$

$$\begin{aligned}(\dot{\hat{\boldsymbol{\omega}}}_{d/c}^{d})^{s}=\boldsymbol{M}_{d}^{-1}*\{&(\hat{\boldsymbol{u}}_{c}^{d}+\hat{\boldsymbol{u}}_{g}^{d}+\hat{\boldsymbol{d}})-(\hat{\boldsymbol{\omega}}_{d/c}^{d}+\hat{\boldsymbol{\omega}}_{c/I}^{d})\times(\boldsymbol{M}_{d}*((\hat{\boldsymbol{\omega}}_{d/c}^{d})^{s}+(\hat{\boldsymbol{\omega}}_{c/I}^{d})^{s}))-\\&\boldsymbol{M}_{d}*(\hat{\boldsymbol{q}}_{d/c}\otimes\dot{\hat{\boldsymbol{\omega}}}_{c/I}^{c}\otimes\hat{\boldsymbol{q}}_{d/c}^{*})^{s}-\boldsymbol{M}_{d}*(\hat{\boldsymbol{\omega}}_{c/I}^{d}\times\hat{\boldsymbol{\omega}}_{d/c}^{d})^{s}\}\end{aligned} \tag{5.95}$$

式中：变量 $\hat{\boldsymbol{\omega}}_{c/I}^{c}$、$\dot{\hat{\boldsymbol{\omega}}}_{c/I}^{c}$ 和 $\hat{\boldsymbol{\omega}}_{c/I}^{d}$ 计算公式为

$$\dot{\hat{\boldsymbol{q}}}_{c/I}=\frac{1}{2}\hat{\boldsymbol{\omega}}^{c}_{c/I}\otimes\hat{\boldsymbol{q}}_{c/I} \tag{5.96}$$

$$(\dot{\hat{\boldsymbol{\omega}}}^{c}_{c/I})^{s}=\boldsymbol{M}_c^{-1}*\{-\hat{\boldsymbol{\omega}}^{c}_{c/I}\times(\boldsymbol{M}_c*(\hat{\boldsymbol{\omega}}^{c}_{c/I})^{s})+\hat{\boldsymbol{u}}^{c}_{g}\} \tag{5.97}$$

$$\hat{\boldsymbol{\omega}}^{d}_{c/I}=\hat{\boldsymbol{q}}_{d/c}\otimes\hat{\boldsymbol{\omega}}^{c}_{c/I}\otimes\hat{\boldsymbol{q}}^{*}_{d/c} \tag{5.98}$$

假设初始时刻主星的姿态角速度为 $\boldsymbol{\omega}^{c}_{c/I}(t_0)=[0,0.0011,-0.0011]^{\mathrm{T}}\mathrm{rad/s}$，相对于主星 LVLH 系的姿态四元数为 $\boldsymbol{q}_{c/H}(t_0)=[0.4\quad -0.2\quad 0.4\quad 0.8]^{\mathrm{T}}$，可依据式（2.29）计算得到初始相对姿态四元数 $\boldsymbol{q}_{d/c}(t_0)$；由于主星的 LVLH 系固连在主星质心上，所以根据前述知识可知 $\boldsymbol{r}^{H}_{c/I}=[r_c\quad 0\quad 0]^{\mathrm{T}}$，$\dot{\boldsymbol{r}}^{H}_{c/I}=[\dot{r}_c\quad 0\quad 0]^{\mathrm{T}}$，$\boldsymbol{\omega}^{H}_{H/I}=[0\quad 0\quad \dot{\theta}]^{\mathrm{T}}$。因此，初始的对偶四元数 $\hat{\boldsymbol{q}}_{c/I}$ 和旋量 $\hat{\boldsymbol{\omega}}^{c}_{c/I}$ 计算公式为

$$\begin{aligned}\hat{\boldsymbol{q}}_{c/I}&=\boldsymbol{q}_{c/I}+\varepsilon\,\frac{1}{2}\boldsymbol{r}^{c}_{c/I}\otimes\boldsymbol{q}_{c/I}\\&=\boldsymbol{q}_{c/I}+\varepsilon\,\frac{1}{2}(\boldsymbol{A}^{c}_{H}(\boldsymbol{q}_{c/H})\boldsymbol{r}^{H}_{c/I})\otimes\boldsymbol{q}_{c/I}\end{aligned} \tag{5.99}$$

$$\begin{aligned}\hat{\boldsymbol{\omega}}^{c}_{c/I}&=\boldsymbol{\omega}^{c}_{c/I}+\varepsilon(\boldsymbol{A}^{c}_{I}\,\dot{\boldsymbol{r}}^{I}_{c/I})\\&=\boldsymbol{\omega}^{c}_{c/I}+\varepsilon(\boldsymbol{A}^{c}_{H}(\boldsymbol{q}_{c/H})\boldsymbol{A}^{H}_{I}\,\dot{\boldsymbol{r}}^{I}_{c/I})\\&=\boldsymbol{\omega}^{c}_{c/I}+\varepsilon(\boldsymbol{A}^{c}_{H}(\boldsymbol{q}_{c/H})(\dot{\boldsymbol{r}}^{H}_{c/I}+\boldsymbol{\omega}^{H}_{H/I}\times\boldsymbol{r}^{H}_{c/I}))\end{aligned} \tag{5.100}$$

式中：$\boldsymbol{q}_{c/I}=\boldsymbol{q}_{c/H}\otimes\boldsymbol{q}_{H/I}$，四元数 $\boldsymbol{q}_{H/I}$ 的计算方法已在前一节中进行了介绍。利用对偶惯性矩阵 $\boldsymbol{M}_c$ 和对偶四元数 $\hat{\boldsymbol{q}}_{c/I}$，可实时计算式（5.97）中 $\hat{\boldsymbol{u}}^{c}_{g}$ 的大小。

对于初始的对偶四元数 $\hat{\boldsymbol{q}}_{d/c}$ 以及旋量 $\hat{\boldsymbol{\omega}}^{d}_{d/c}$，其计算公式为

$$\begin{aligned}\hat{\boldsymbol{q}}_{d/c}&=\boldsymbol{q}_{d/c}+\varepsilon\,\frac{1}{2}\boldsymbol{p}^{d}_{d/c}\otimes\boldsymbol{q}_{d/c}\\&=\boldsymbol{q}_{d/c}+\varepsilon\,\frac{1}{2}(\boldsymbol{A}^{d}_{H}(\boldsymbol{q}_{d/H})\boldsymbol{p}^{H}_{d/c})\otimes\boldsymbol{q}_{d/c}\end{aligned} \tag{5.101}$$

$$\begin{aligned}\hat{\boldsymbol{\omega}}^{d}_{d/c}&=\boldsymbol{\omega}^{d}_{d/c}+\varepsilon(\boldsymbol{A}^{d}_{c}(\boldsymbol{q}_{d/c})\dot{\boldsymbol{p}}^{c}_{d/c})\\&=\boldsymbol{\omega}^{d}_{d/c}+\varepsilon(\boldsymbol{A}^{d}_{H}(\boldsymbol{q}_{d/H})\boldsymbol{A}^{H}_{c}(\boldsymbol{q}_{H/c})\dot{\boldsymbol{p}}^{c}_{d/c})\\&=\boldsymbol{\omega}^{d}_{d/c}+\varepsilon(\boldsymbol{A}^{d}_{H}(\boldsymbol{q}_{d/H})(\dot{\boldsymbol{p}}^{H}_{d/c}+\boldsymbol{\omega}^{H}_{H/c}\times\boldsymbol{p}^{H}_{d/c}))\end{aligned} \tag{5.102}$$

式中：角速度 $\boldsymbol{\omega}^{d}_{d/c}$ 和 $\boldsymbol{\omega}^{H}_{H/c}$ 计算公式为

$$\boldsymbol{\omega}^{d}_{d/c}=\boldsymbol{\omega}^{d}_{d/I}-\boldsymbol{A}^{d}_{c}(\boldsymbol{q}_{d/c})\boldsymbol{\omega}^{c}_{c/I} \tag{5.103}$$

$$\boldsymbol{\omega}^{H}_{H/c}=\boldsymbol{\omega}^{H}_{H/I}-\boldsymbol{A}^{H}_{c}(\boldsymbol{q}^{-1}_{c/H})\boldsymbol{\omega}^{c}_{c/I} \tag{5.104}$$

类似地，结合第 4 章中的相对导航方法以及式（5.97）~式（5.104），亦可实时地为式（5.63）或式（5.75）提供所需参数值，进行导航控制一体化仿真。

仿真参数设置如前所述，采用式（5.75）中的姿轨一体化控制律，控制参数选取为 $\omega_r=\omega_\theta=1/30$，$\zeta_r=\zeta_\theta=1.0$，并利用 5.2.3 节中的控制分配算法实现全推力器姿轨一体化控制，仿真时间为 300s。仿真结果如图 5.10～图 5.16 所示。

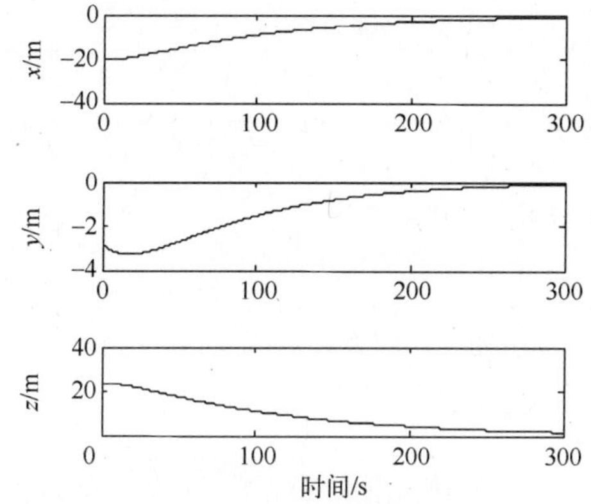

图 5.10　相对位置 $\boldsymbol{p}_{d/c}^{c}$

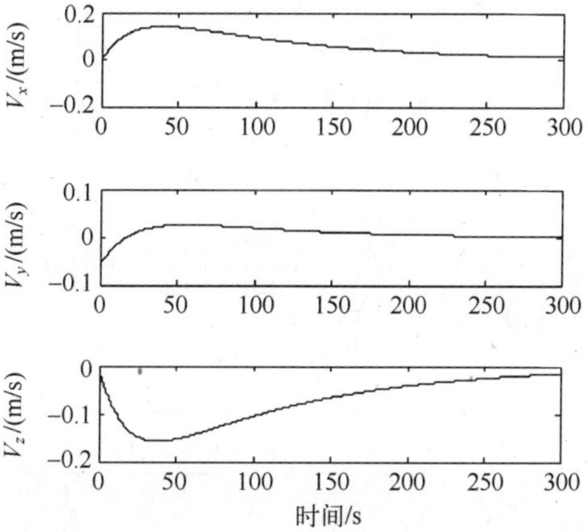

图 5.11　相对速度 $\dot{\boldsymbol{p}}_{d/c}^{c}$

图 5.10 和图 5.11 分别给出了在主星体系中的相对位置和相对速度曲线。可以看出，相对位置和速度指向变化平稳，且逐渐收敛到零。图 5.12 和图 5.13 分别给出了相对姿态四元数 $\boldsymbol{q}_{d/c}$ 和相对姿态角速度 $\boldsymbol{\omega}_{d/c}^{d}$ 的变化曲线。图 5.14～图 5.16 分别给出了控制过程中 1～12 号推力器的控制指令序列，各推力器控制输出均控制在 0.1N 以内。

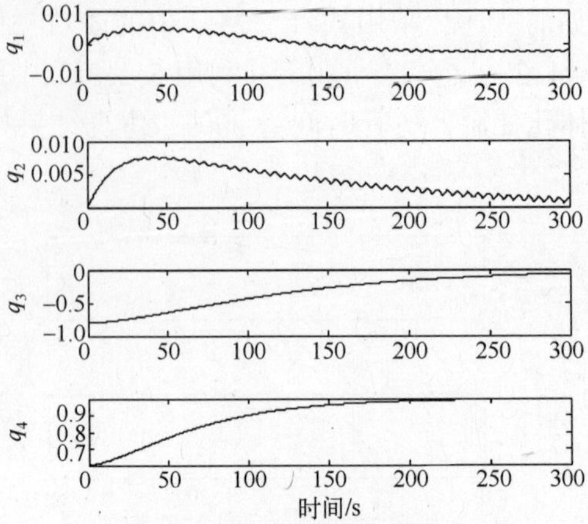

图 5.12 相对姿态四元数 $\boldsymbol{q}_{d/c}$

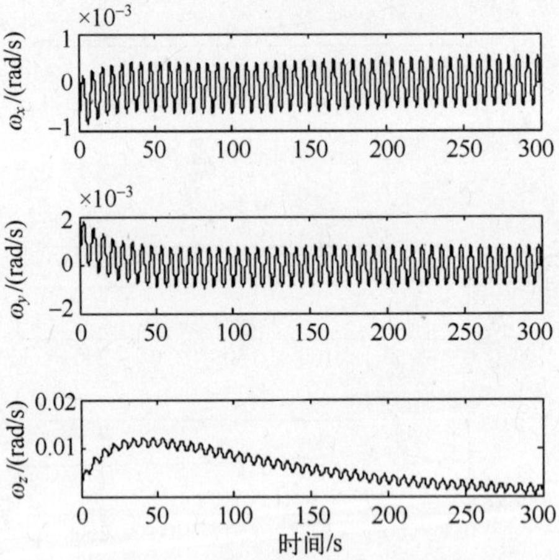

图 5.13 相对姿态角速度 $\boldsymbol{\omega}_{d/c}^{d}$

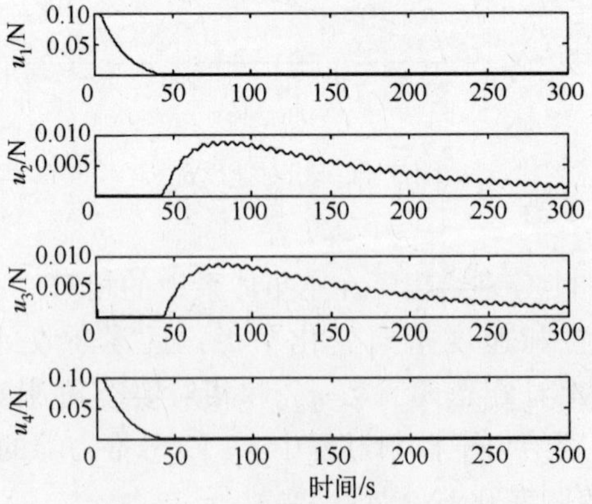

图 5.14 1~4 号推力器指令变化

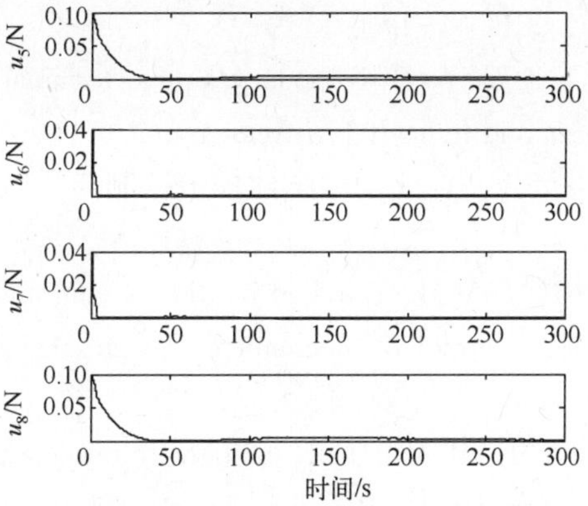

图 5.15　5~8 号推力器指令变化

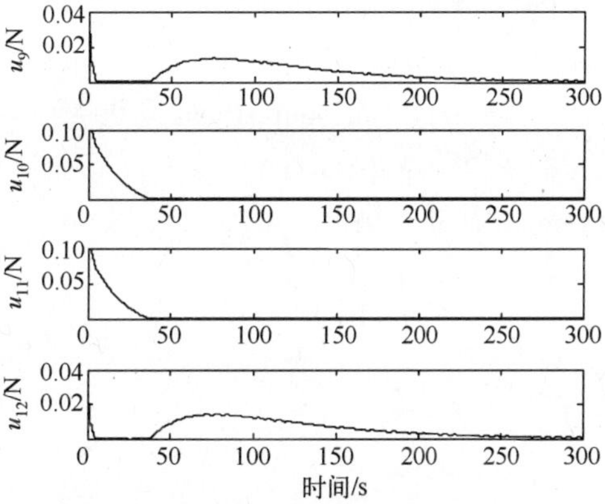

图 5.16　9~12 号推力器指令变化

参 考 文 献

[1] WU Y X, HU X P, HU D W, et al. Strapdown inertial navigation system algorithms based on dual quaternions [J]. IEEE Aerospace and Electronic Systems, 2005, 41 (1): 110-132.

[2] 武元新. 对偶四元数导航算法与非线性高斯滤波研究 [D]. 长沙: 国防科学技术大学, 2005.

[3] ZHANG F, DUAN G R. Robust integrated translation and rotation finite-time maneuver of a rigid spacecraft based on dual quaternion [C] //AIAA Guidance, Navigation, and Control Conference, August 2011, Portland, Oregon.

USA: AIAA Inc, 2011: 1-17.
[4] WANG J Y, SUN Z W. 6-DOF robust adaptive terminal sliding mode control for spacecraft formation flying [J]. Acta Astronautica, 2012, 73: 76-87.
[5] 王剑颖. 航天器姿轨一体化动力学建模、控制与导航方法研究 [D]. 哈尔滨: 哈尔滨工业大学, 2013.
[6] BRODSKY V, SHOHAM M. The dual inertia operator and its application to robot dynamics [J]. Journal of Mechanical Design, 1994, 116 (4): 1189-1195.
[7] BRODSKY V, SHOHAM M. Dual numbers representation of rigid body dynamics [J]. Mechanism and Machine Theory, 1999, 34 (5): 693-718.
[8] WERTZ J R. Spacecraft attitude determination and control [M]. Dordrecht: Kluwer Academic Publishers, 1978.
[9] SIDI M J. Spacecraft dynamics and control: a practical engineering approach [M]. New York: Cambridge University Press, 1997.
[10] SLOTINE J J, LI W P. Applied nonlinear control [M]. Upper Saddle River, N J: Prentice Hall, 1991.

第 6 章　空间高精度姿态机动控制

航天器在轨运行期间需要进行大量的姿态机动来完成预定的观测等定向任务。本章针对采用 SGCMG 系统作为执行机构的航天器大角度姿态机动任务，开展两种空间高精度姿态机动控制方法研究：①基于路径规划和线性二次调节（LQR）反馈控制相结合的姿态机动控制方法；②模型误差预测控制方法。

首先，提出一种基于路径规划和 LQR 反馈控制相结合的姿态机动控制方法。针对固定时间能量最优和准时间最优两种情形，采用 Radau 伪谱法优化机动路径并回避奇异状态，将最优控制量作为参考输入，并利用基于 LQR 的最优反馈控制方法消除初始偏差、模型不确定性以及外界扰动等影响。该方法将航天器和 SGCMG 系统作为整体规划姿态运动，能够有效避免奇异状态。

其次，针对模型参数不确定性以及外界干扰等情形，进一步研究模型误差预测控制（Model Error Predictive Control，MEPC）方法。MEPC 是一种鲁棒控制方法，该方法包括一个标称控制器和一个模型误差估计器，利用预测滤波方法来确定模型误差，从而可以有效地补偿由参数模型误差和外界扰动引起的误差。

6.1　基于路径规划的最优姿态快速机动控制律设计与仿真

6.1.1　最优姿态轨迹规划问题

6.1.1.1　拉格朗日型最优控制问题

考虑如下一般拉格朗日型最优控制问题，优化指标函数，即

$$J=\int_{t_0}^{t_f}\mathcal{L}(\boldsymbol{x}(t),\boldsymbol{u}(t),t)\,\mathrm{d}t \tag{6.1}$$

且满足如下动力学约束、边界条件和过程约束：

$$\frac{\mathrm{d}\boldsymbol{x}}{\mathrm{d}t}=\boldsymbol{f}(\boldsymbol{x}(t),\boldsymbol{u}(t),t) \tag{6.2}$$

$$\boldsymbol{\phi}(\boldsymbol{x}(t_0),t_0,\boldsymbol{x}(t_f),t_f)=\boldsymbol{0} \tag{6.3}$$

$$\boldsymbol{P}(\boldsymbol{x}(t),\boldsymbol{u}(t),t)\leqslant\boldsymbol{0} \tag{6.4}$$

式中：$\boldsymbol{x}(t)\in\mathbb{R}^n$ 和 $\boldsymbol{u}(t)\in\mathbb{R}^m$ 分别为待求解的状态变量和控制变量；t_0 和 t_f 分别为初始和终端时刻。

6.1.1.2 航天器姿态机动规划问题

本章采用2.4.2节中的金字塔构型SGCMG系统作为姿态执行机构，对航天器和4个SGCMG构成的整体系统进行规划的方法，在航天器姿态机动前规划出一条合理的路径，以减小在整个机动过程中SGCMG系统陷入奇异的可能性。

状态变量选取为

$$\begin{aligned}\boldsymbol{x}&=[\boldsymbol{q}^{\mathrm{T}}\quad\boldsymbol{\omega}^{\mathrm{T}}\quad\boldsymbol{\delta}^{\mathrm{T}}]^{\mathrm{T}}\\&=[q_1\quad q_2\quad q_3\quad q_4\quad\omega_1\quad\omega_2\quad\omega_3\quad\delta_1\quad\delta_2\quad\delta_3\quad\delta_4]^{\mathrm{T}}\end{aligned} \tag{6.5}$$

控制变量为 $\boldsymbol{u}=[u_1\quad u_2\quad u_3\quad u_4]^{\mathrm{T}}$。结合航天器姿态运动学和动力学模型以及2.4.2节中的控制力矩陀螺模型，可得如下状态方程约束：

$$\begin{cases}\dot{q}_1=0.5(q_4\omega_1-q_3\omega_2+q_2\omega_3)\\\dot{q}_2=0.5(q_3\omega_1+q_4\omega_2-q_1\omega_3)\\\dot{q}_3=0.5(-q_2\omega_1+q_1\omega_2+q_4\omega_3)\\\dot{q}_4=0.5(-q_1\omega_1-q_2\omega_2-q_3\omega_3)\\\dot{\omega}_1=J_{11}^{-1}[(J_{22}-J_{33})\omega_2\omega_3-(\omega_2h_3-\omega_3h_2)-\dot{h}_1]\\\dot{\omega}_2=J_{22}^{-1}[(J_{33}-J_{11})\omega_1\omega_3-(\omega_3h_1-\omega_1h_3)-\dot{h}_2]\\\dot{\omega}_3=J_{33}^{-1}[(J_{11}-J_{22})\omega_1\omega_2-(\omega_1h_2-\omega_2h_1)-\dot{h}_3]\\\dot{\delta}_i=u_i,i=1,2,3,4\end{cases} \tag{6.6}$$

式中：J_{11}，J_{22} 和 J_{33} 分别为航天器的主惯量；$\dot{h}_1$，$\dot{h}_2$ 和 $\dot{h}_3$ 为角动量 $\dot{\boldsymbol{h}}$ 的三分量。注意到方程组（6.6）是用来规划理想姿态机动轨迹，而干扰力矩以及惯量参数不确定性等情形将会在后续设计的闭环控制器中考虑到。

路径约束主要考虑航天器姿态机动角速度受限、SGCMG框架角受限、框架角速度受限以及奇异量度受限等问题：

$$\begin{aligned}&|\omega_i|\leqslant\omega_{i\max},\quad i=1,2,3\\&|\delta_i|\leqslant\delta_{i\max},\quad i=1,2,3,4\\&|u_i|\leqslant\dot{\delta}_{i\max},\quad i=1,2,3,4\\&D\geqslant D_{\min}\end{aligned} \tag{6.7}$$

边界条件为

$$\boldsymbol{x}(t_0)=\boldsymbol{x}_0,\boldsymbol{x}(t_f)=\boldsymbol{x}_f \tag{6.8}$$

航天器大角度姿态机动要求在满足方程（6.7）约束条件下从初始状态 $\boldsymbol{x}_0$ 机动到期望状态 $\boldsymbol{x}_f$，性能优化指标可以选取固定时间能量最优，或者准时间最优，即如下两种性能优化指标：

$$J=\min_u \int_{t_0}^{t_f} \boldsymbol{u}^{\mathrm{T}}\boldsymbol{u}\mathrm{d}t \tag{6.9}$$

$$J=\min_u \int_{t_0}^{t_f} (1+\boldsymbol{u}^{\mathrm{T}}\boldsymbol{u})\mathrm{d}t \tag{6.10}$$

6.1.2 Radau 伪谱法

伪谱法利用全局插值多项式的有限基，在一系列离散点上近似状态变量和控制变量，将最优控制问题转化为非线性规划问题来求解。微分方程中状态的时间导数通过插值多项式的导数来近似，进而在一组配点上将微分方程转化为代数方程。常见的伪谱方法包括 Legendre 伪谱法（Legendre Pseudospectral Method，LPM）[1]、高斯伪谱法（Gauss Pseudospectral Method，GPM）[2-3]以及 Radau 伪谱法（Radau Pseudospectral Method，RPM）[4-6]。由于 Radau 伪谱法具有较高的计算精度以及计算效率[7]，本章采用 Radau 伪谱法作为优化算法。

Radau 伪谱法将最优控制问题离散转录为非线性规划问题，通过如下线性变换可将时间间隔区间 $t\in[t_0,t_f]$ 转化至区间 $\boldsymbol{\tau}\in[-1,1]$：

$$t=\frac{t_f-t_0}{2}\boldsymbol{\tau}+\frac{t_f+t_0}{2} \tag{6.11}$$

因此，方程（6.1）~方程（6.4）的最优控制问题可以在新的计算时间域中改写为

$$J=\frac{t_f-t_0}{2}\int_{-1}^{1}\mathcal{L}(\boldsymbol{x}(\boldsymbol{\tau}),\boldsymbol{u}(\boldsymbol{\tau}),\boldsymbol{\tau})\mathrm{d}\boldsymbol{\tau} \tag{6.12}$$

满足如下约束条件：

$$\frac{\mathrm{d}\boldsymbol{x}}{\mathrm{d}\boldsymbol{\tau}}=\frac{t_f-t_0}{2}\boldsymbol{f}(\boldsymbol{x}(\boldsymbol{\tau}),\boldsymbol{u}(\boldsymbol{\tau}),\boldsymbol{\tau};t_0,t_f) \tag{6.13}$$

$$\boldsymbol{\phi}(\boldsymbol{x}(-1),t_0,\boldsymbol{x}(1),t_f)=\boldsymbol{0} \tag{6.14}$$

$$\boldsymbol{P}(\boldsymbol{x}(\boldsymbol{\tau}),\boldsymbol{u}(\boldsymbol{\tau}),\boldsymbol{\tau};t_0,t_f)\leqslant\boldsymbol{0} \tag{6.15}$$

在 RPM 中，考虑 LGR 配点 $-1=\tau_1<\cdots<\tau_N<+1$ 以及端点 $\boldsymbol{\tau}_{N+1}=1$。状态矢量可以用 N 阶拉格朗日插值多项式拟合状态量，即

$$\boldsymbol{x}(\boldsymbol{\tau}) \approx \boldsymbol{X}(\boldsymbol{\tau}) = \sum_{i=1}^{N+1} \boldsymbol{X}(\boldsymbol{\tau}_i) L_i(\boldsymbol{\tau}) \tag{6.16}$$

式中：$L_i(\boldsymbol{\tau})$定义为

$$L_i(\boldsymbol{\tau}) = \prod_{j=1,j\neq i}^{N+1} \frac{\boldsymbol{\tau} - \boldsymbol{\tau}_j}{\boldsymbol{\tau}_i - \boldsymbol{\tau}_j}, i = 1,2,\cdots,N+1 \tag{6.17}$$

对式（6.16）求微分可得到状态量$\boldsymbol{x}(\boldsymbol{\tau})$在 LGR 点$\tau_k(k=1,2,\cdots,N)$处的导数：

$$\dot{\boldsymbol{x}}(\boldsymbol{\tau}_k) \approx \dot{\boldsymbol{X}}(\boldsymbol{\tau}_k) = \sum_{i=1}^{N+1} \boldsymbol{X}(\boldsymbol{\tau}_i) \dot{L}_i(\boldsymbol{\tau}_k) = \sum_{i=1}^{N+1} D_{ki}\boldsymbol{X}(\boldsymbol{\tau}_i) \tag{6.18}$$

式中：D_{ki}是 Radau 伪谱微分系数矩阵$\boldsymbol{D} \in \mathbb{R}^{N\times(N+1)}$的元素。利用微分系数矩阵可将式（6.13）的动态约束条件转化为如下代数约束：

$$\sum_{i=1}^{N+1} \boldsymbol{D}_{ki}\boldsymbol{X}_i - \frac{t_f - t_0}{2}\boldsymbol{f}(\boldsymbol{X}_k, \boldsymbol{U}_k, \boldsymbol{\tau}_k; t_0, t_f) = \boldsymbol{0}, \quad k = 1,2,\cdots,N \tag{6.19}$$

其中，$\boldsymbol{X}_k \equiv \boldsymbol{X}(\boldsymbol{\tau}_k)$，$\boldsymbol{U}_k \equiv \boldsymbol{U}(\boldsymbol{\tau}_k)$。注意到动态约束条件仅在$N$个 LGR 点采集，而状态除了在$N$个 LGR 点加上端点$\boldsymbol{\tau}_{N+1}=1$。类似地，指标函数（6.12）可以用 Gauss-Radau 求积函数近似成

$$J = \frac{t_f - t_0}{2} \sum_{k=1}^{N} w_k \mathcal{L}(\boldsymbol{X}_k, \boldsymbol{U}_k, \boldsymbol{\tau}_k; t_0, t_f) \tag{6.20}$$

其中，w_k为 Gauss 权重系数。最后，离散的边界条件和路径约束可以表示成

$$\boldsymbol{\phi}(\boldsymbol{X}_1, t_0, \boldsymbol{X}_{N+1}, t_f) = \boldsymbol{0} \tag{6.21}$$

$$\boldsymbol{P}(\boldsymbol{X}_k, \boldsymbol{U}_k, \tau_k; t_0, t_f) \leqslant \boldsymbol{0}, \quad k=1,2,\cdots,N \tag{6.22}$$

因此，NLP 问题可以表示成：确定状态变量$\boldsymbol{X}_k$和控制变量$\boldsymbol{U}_k$，使得指标函数（6.20）在满足代数约束式（6.19）、式（6.21）和式（6.22）条件下取最小值。

6.1.3　基于 LQR 的轨迹跟踪控制

一般情况下，采用伪谱法求解的最优路径已经具有较高的精度，但是采用开环控制方法通常不具有对初始条件和扰动的鲁棒性。例如，由于离线规划需要已知初始框架角$\boldsymbol{\delta}$，当利用规划好的姿态轨迹进行控制时，初始框架角通常已经发生改变；另外，由于对 SGCMG 建模所进行的简化，忽略了框架角动量、动量轮框架轴方向和控制力矩方向角动量以及环境扰动力矩。这些初始条件不确定和扰动将对开环控制造成较大影响，使得终端姿态偏离期望姿态。因此，需要进行反馈控制，这里采用基于 LQR 的轨迹跟踪控制方法。如图 6.1 所示，控制量由两部分组成：一部分是伪谱法优化得到的参考控制量

$\boldsymbol{u}_{\text{ref}}$，另一部分是反馈控制量 $\delta\boldsymbol{u}$。

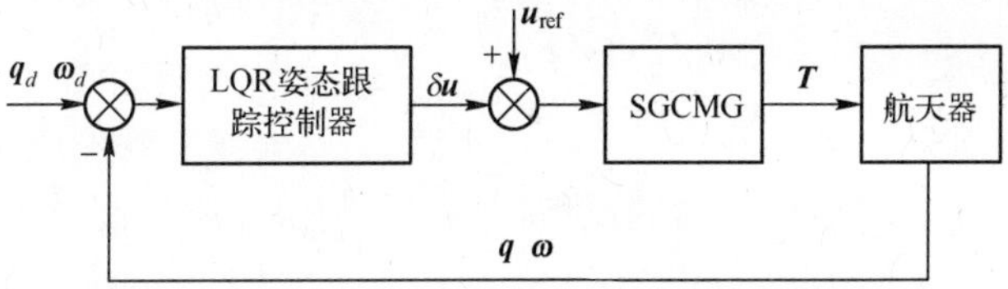

图 6.1　LQR 姿态跟踪控制流程

选取跟踪状态变量 $\boldsymbol{x}=[\boldsymbol{Q}^{\mathrm{T}}\quad\boldsymbol{\omega}^{\mathrm{T}}]^{\mathrm{T}}$，控制变量 $\boldsymbol{u}=\dot{\boldsymbol{\delta}}$，结合航天器姿态动力学和运动学模型，以及 SGCMG 模型，可将非线性系统动力学方程改写为

$$\begin{aligned}\dot{\boldsymbol{x}}&=\boldsymbol{f}(\boldsymbol{x},\boldsymbol{u})\\&=\begin{bmatrix}\dfrac{1}{2}([\boldsymbol{Q}\times]+q_4\boldsymbol{I}_{3\times3})\boldsymbol{\omega}\\-\boldsymbol{J}^{-1}[\boldsymbol{\omega}\times(\boldsymbol{J}\boldsymbol{\omega}+\boldsymbol{h})+\boldsymbol{C}(\boldsymbol{\delta})\boldsymbol{u}]\end{bmatrix}\end{aligned}\tag{6.23}$$

定义相对标称轨迹（最优姿态机动轨迹）的扰动状态 $\delta\boldsymbol{x}=[\delta\boldsymbol{Q}^{\mathrm{T}}\quad\delta\boldsymbol{\omega}^{\mathrm{T}}]^{\mathrm{T}}$ 和扰动控制变量 $\delta\boldsymbol{u}$。对式（6.23）在标称轨迹（最优姿态机动轨迹）附近线性化，可得

$$\delta\dot{\boldsymbol{x}}=\boldsymbol{A}(t)\delta\boldsymbol{x}+\boldsymbol{B}(t)\delta\boldsymbol{u}\tag{6.24}$$

式中：

$$\boldsymbol{A}=\frac{\partial\boldsymbol{f}}{\partial\boldsymbol{x}}=\begin{bmatrix}-\dfrac{1}{2}[\boldsymbol{\omega}\times]&\dfrac{1}{2}([\boldsymbol{Q}\times]+q_4\boldsymbol{I}_{3\times3})\\\boldsymbol{0}_{3\times3}&\boldsymbol{J}^{-1}([(\boldsymbol{J}\boldsymbol{\omega}+\boldsymbol{h})\times]-[\boldsymbol{\omega}\times]\boldsymbol{J})\end{bmatrix}$$

$$\boldsymbol{B}=\begin{bmatrix}\boldsymbol{0}_{3\times4}\\-\boldsymbol{J}^{-1}\boldsymbol{C}(\boldsymbol{\delta})\end{bmatrix}$$

选取二次性能指标

$$J=\frac{1}{2}\int_0^{t_f}(\delta\boldsymbol{x}^{\mathrm{T}}\boldsymbol{Q}\delta\boldsymbol{x}+\delta\boldsymbol{u}^{\mathrm{T}}\boldsymbol{R}\delta\boldsymbol{u})\,\mathrm{d}t\tag{6.25}$$

式中：$\boldsymbol{Q}\in\mathbb{R}^{6\times6}$表示对系统误差的惩罚大小，是半正定矩阵；$\boldsymbol{R}\in\mathbb{R}^{4\times4}$表示对系统控制能量的惩罚，是正定矩阵。

反馈增益矩阵 $\boldsymbol{K}$ 和最优反馈控制律 $\delta\boldsymbol{u}$ 为

$$\boldsymbol{K}=\boldsymbol{R}^{-1}\boldsymbol{B}^{\mathrm{T}}\boldsymbol{P}\tag{6.26}$$

$$\delta \boldsymbol{u} = -\boldsymbol{K}\delta \boldsymbol{x} \tag{6.27}$$

式中：$\boldsymbol{P}$ 为 Riccati 方程的解，即

$$\boldsymbol{PA}+\boldsymbol{AP}-\boldsymbol{PBR}^{-1}\boldsymbol{BP}+\boldsymbol{Q}=\boldsymbol{0} \tag{6.28}$$

结合标称轨迹，通过事先调节权重矩阵 $\boldsymbol{Q}$ 和 $\boldsymbol{R}$，离线生成反馈增益系数矩阵。结合式（6.27）中施加的反馈控制，可得到实际控制量为

$$\boldsymbol{u} = \boldsymbol{u}_{\text{ref}}+\delta \boldsymbol{u} \tag{6.29}$$

6.1.4 仿真实验与分析

航天器惯量张量矩阵取为 $\boldsymbol{J}=\text{diag}(70.75,65.45,75.45)\text{kg}\cdot\text{m}^2$，金字塔构型 SGCMG 系统的安装倾角 β 为 53.13°，每个 SGCMG 角动量大小 $h=1.5\text{N}\cdot\text{m}\cdot\text{s}$。采用高斯伪谱法 MATLAB 软件包 GPOPS（Gauss Pseudospectral Optimization Software）对基于 SGCMG 卫星的大角度姿态机动进行路径规划，对于路径约束条件（6.7），$\omega_{i\max} \leqslant 4°/\text{s}$，$\delta_{i\max}=\pi$，$\dot{\delta}_{i\max}=2\text{rad/s}$，$D_{\min}=0.4$。初始状态 $\boldsymbol{x}_0=[0,0,0,1,0,0,0,0,0,0,0,0]^{\text{T}}$，考虑绕滚动轴的 45°大角度机动，初始姿态为零，SGCMG 系统终端框架角 $\boldsymbol{\delta}_f$ 不作约束，终端三轴姿态角速度皆为 0，则此时的终端状态 $\boldsymbol{x}_f=[0.3827,0,0,0.9239,0,0,0,\delta_{1f},\delta_{2f},\delta_{3f},\delta_{4f}]^{\text{T}}$。

6.1.4.1 最优姿态机动轨迹

本节采用自适应高斯伪谱法进行姿态机动轨迹优化。整个大角度机动轨迹优化计算在 CPU 为 Pentium（R）Dual Core、主频为 2.93GHz，内存为 2GB 的 PC 机上运行，算法在 MATLAB R2010a 环境下采用 GPOPS 软件包编程实现。GPOPS 是由 Rao 等[8]开发的，采用自适应高斯和 Radau 伪谱法在 MATLAB 环境下求解复杂最优控制问题的免费软件包。

1. 情形一：固定时间能量最优

考虑如下能量优化指标：

$$J = \min_{u} \int_{t_0}^{t_f} \boldsymbol{u}^{\text{T}}\boldsymbol{u}\,\text{d}t \tag{6.30}$$

选取终端时间 $t_f=20\text{s}$，仿真结果如图 6.2～图 6.7 所示。仿真结果表明，所求得的优化轨迹变化平缓，框架角、框架角速度和奇异量度 D 始终满足约束条件，性能指标 $J=4.16$。

2. 情形二：准时间最优

兼顾考虑时间最优的要求，可考虑如下准时间最优性能指标：

$$J = \min_{u} \int_{t_0}^{t_f} (1 + \boldsymbol{u}^{\text{T}}\boldsymbol{u})\,\text{d}t \tag{6.31}$$

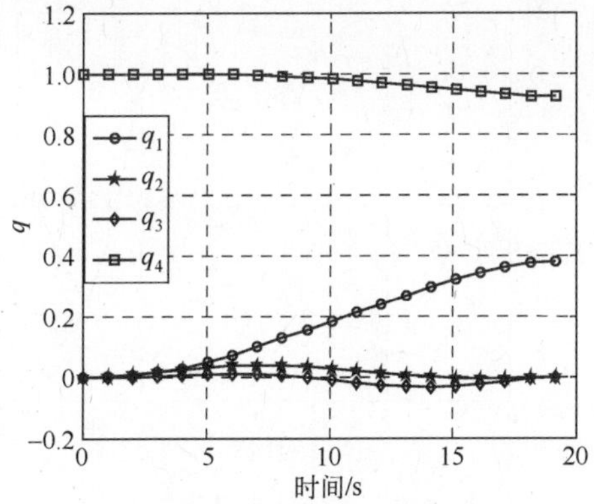

图 6.2　四元数最优轨迹

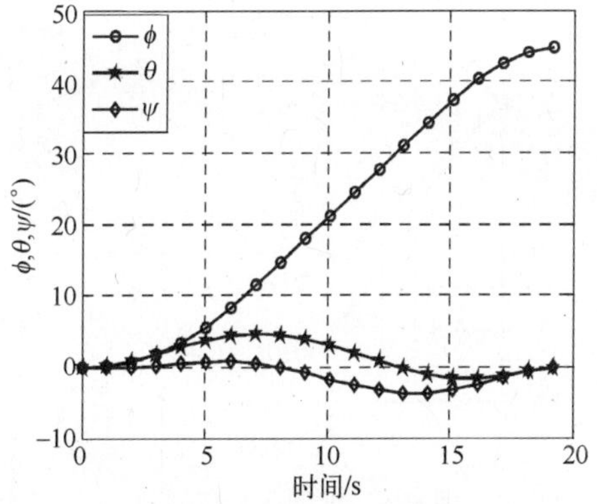

图 6.3　姿态角最优轨迹

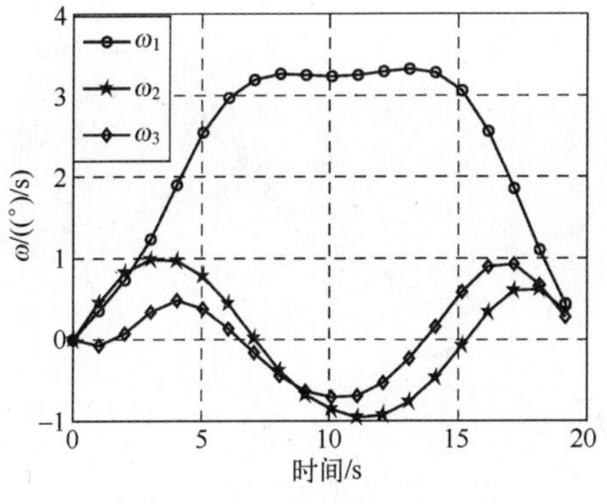

图 6.4　姿态角速度最优轨迹

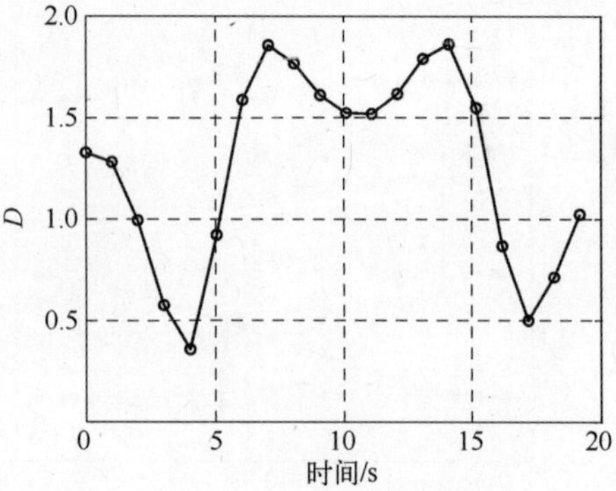

图 6.5　奇异量度最优轨迹

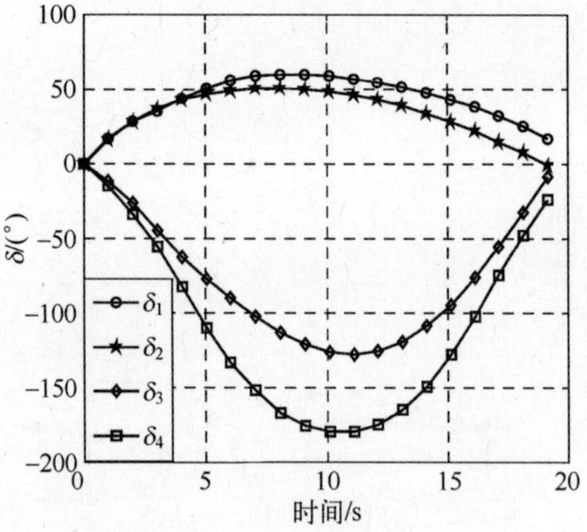

图 6.6　框架角最优轨迹

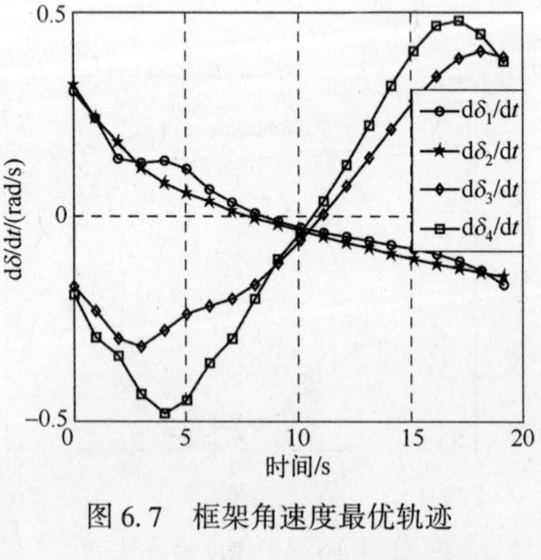

图 6.7　框架角速度最优轨迹

仿真结果如图 6.8～图 6.13 所示。求得的优化轨迹变化平缓，框架角、框架角速度和奇异量度 D 始终满足约束条件。仿真结果表明，在满足各类约束条件下，航天器实现绕滚动轴机动 45°仅需耗时 16s，性能指标 $J=22.34$，其中能量优化指标 $J_u=6.34$，可以看出，相比前一算例消耗了更多能量，但节省了时间，与理论相符。

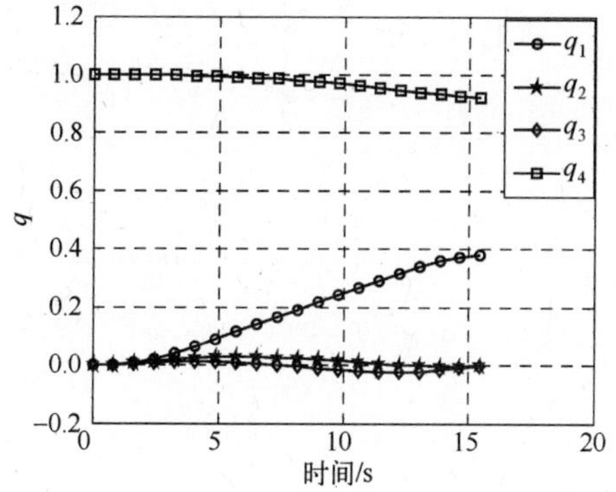

图 6.8　四元数最优轨迹

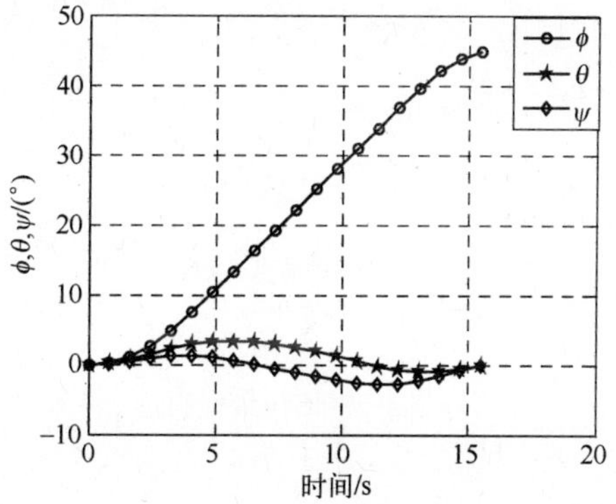

图 6.9　姿态角最优轨迹

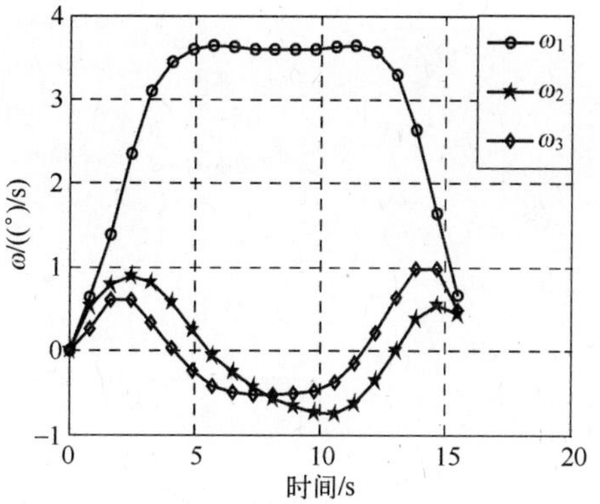

图 6.10　姿态角速度最优轨迹

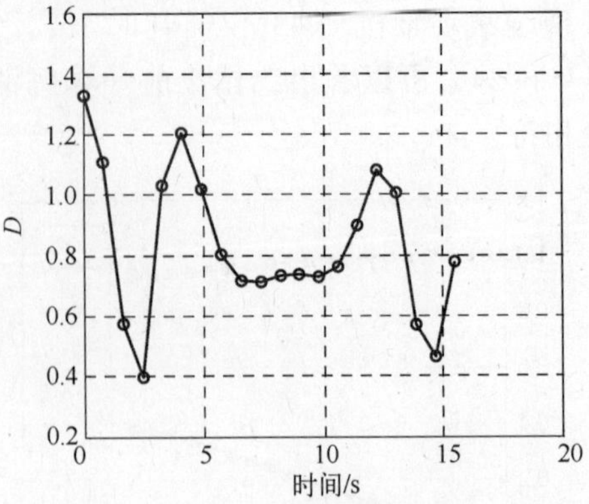

图 6.11　奇异量度最优轨迹

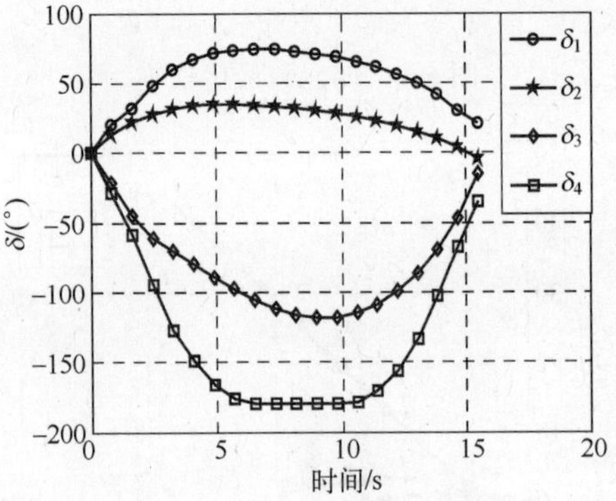

图 6.12　框架角最优轨迹

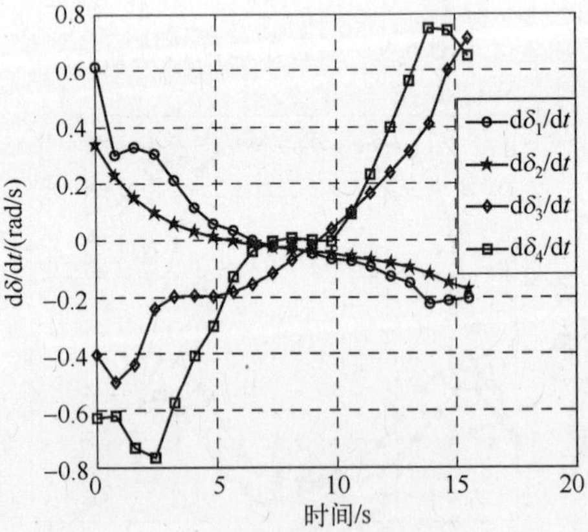

图 6.13　框架角速度最优轨迹

可以看出，该方法综合考虑了实际工程应用中存在的各种约束条件，给出了采用 SGCMG 系统作为执行机构的航天器大角度机动过程中的开环无奇异系统轨迹。该方法能够有效回避奇异状态，不会引入力矩误差，而且综合考虑了 SGCMG 框架角受限、框架角速度受限、奇异量度受限、航天器机动角速度受限以及初始和终端状态受限等约束条件，更贴近工程实际。与其他离线优化方法相比，本节提出的方法无须提供航天器大角度机动过程中的期望力矩变化轨迹的信息，只需提供航天器的始末状态以及各种约束条件。在 MATLAB 环境下，该算法可以在 25s 内快速给出实现航天器大角度机动任务的无奇异光滑轨迹。

6.1.4.2　闭环控制

如前所述，尽管采用伪谱法求解的最优路径已经具有较高的精度，但是采用开环控制方法通常不具有对初始条件和扰动的鲁棒性。在实际工程应用当中，总会存在一些不确定性和扰动，这些因素将会对开环控制方法造成较大影响，使得终端姿态偏离期望姿态。因此，本节采用基于 LQR 的最优反馈控制方法消除不确定性和扰动的影响。

在前面的仿真参数基础上，设置以下初始不确定性和扰动参数：

（1）实际惯量张量矩阵 $\boldsymbol{J}=\bar{\boldsymbol{J}}+\Delta\boldsymbol{J}$，标称惯量张量阵 $\bar{\boldsymbol{J}}$ 和惯量张量不确定部分 $\Delta\boldsymbol{J}$ 分别为

$$\bar{\boldsymbol{J}}=\begin{bmatrix}70.75 & 0 & 0\\ 0 & 65.45 & 0\\ 0 & 0 & 75.45\end{bmatrix}\mathrm{kg}\cdot\mathrm{m}^2,\quad \Delta\boldsymbol{J}=\begin{bmatrix}7 & 0 & 0\\ 0 & 6 & 0\\ 0 & 0 & 7\end{bmatrix}\mathrm{kg}\cdot\mathrm{m}^2$$

（2）初始角速度偏差 $\boldsymbol{\omega}(0)=[0.001\quad 0.001\quad 0.001]^{\mathrm{T}}\mathrm{rad/s}$。

（3）外部干扰力矩为

$$\boldsymbol{d}=5\times10^{-4}\begin{bmatrix}\sin(10^{-4}t)\\ 1\\ \cos(10^{-4}t)\end{bmatrix}\mathrm{N}\cdot\mathrm{m}$$

以固定时间能量最优姿态机动轨迹为例，本节在伪谱法求解的最优路径基础上，采用基于 LQR 的最优反馈控制方法消除不确定性和扰动的影响。采用 6.1.3 节中的方法，把前一节优化得到的最优姿态机动轨迹定义为标称轨迹，离线生成 LQR 反馈增益矩阵。表 6.1 对比了开环控制与闭环控制终端状态误差。

表 6.1 开环控制与闭环控制终端误差对比（固定时间能量最优）

控制方法		$\varphi/(°)$	$\psi/(°)$	$\gamma/(°)$	$\omega_x/((°)/s)$	$\omega_y/((°)/s)$	$\omega_z/((°)/s)$	J_u
开环控制		−2.89	2.11	0.28	0.05	0.10	0.01	4.16
闭环控制	$\boldsymbol{Q}=1\times10^3\boldsymbol{I}_{6\times6}$ $\boldsymbol{R}=\boldsymbol{I}_{4\times4}$	−0.19	0.01	0.05	0.05	−0.01	7×10^{-4}	4.64
	$\boldsymbol{Q}=1\times10^3\boldsymbol{I}_{6\times6}$ $\boldsymbol{R}=0.1\boldsymbol{I}_{4\times4}$	−0.01	−0.01	0.004	1×10^{-3}	1×10^{-3}	2×10^{-4}	4.72

可以看出，通过反馈控制，姿态角和角速度控制精度相比开环控制得到了较大改善。调小权重矩阵 $\boldsymbol{R}$，控制精度有一定提高，满足精度指标要求，但也消耗了更多能量。图 6.14～图 6.18 给出了对应于权重矩阵 $\boldsymbol{Q}=1\times10^3\boldsymbol{I}_{6\times6}$，$\boldsymbol{R}=0.1\boldsymbol{I}_{4\times4}$ 情形下的闭环控制仿真结果。

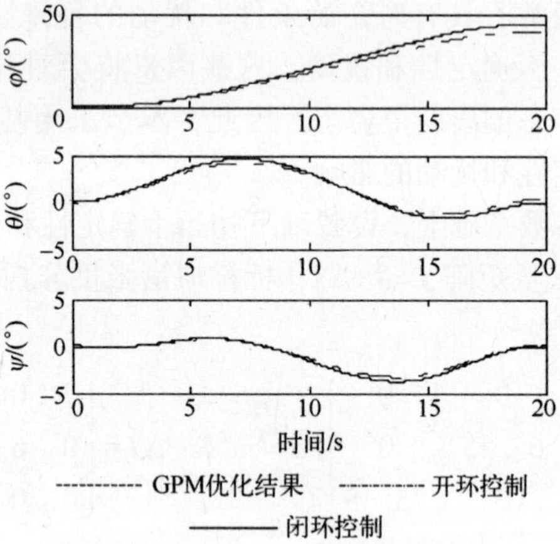

图 6.14 姿态角变化曲线

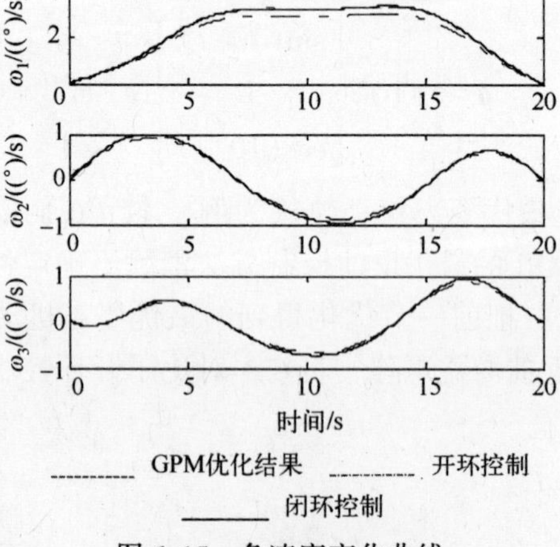

图 6.15 角速度变化曲线

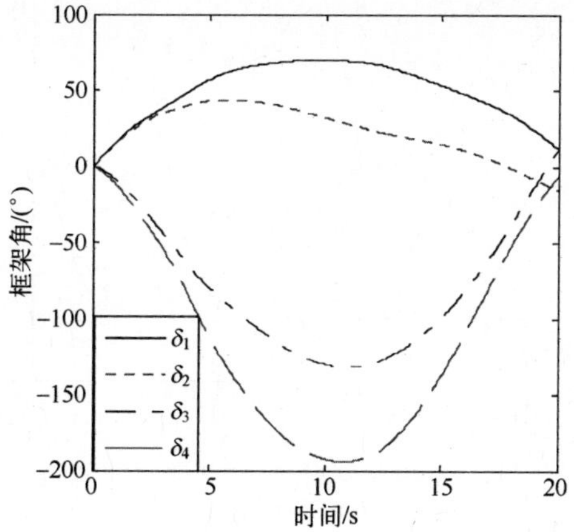

图 6.16　框架角变化曲线

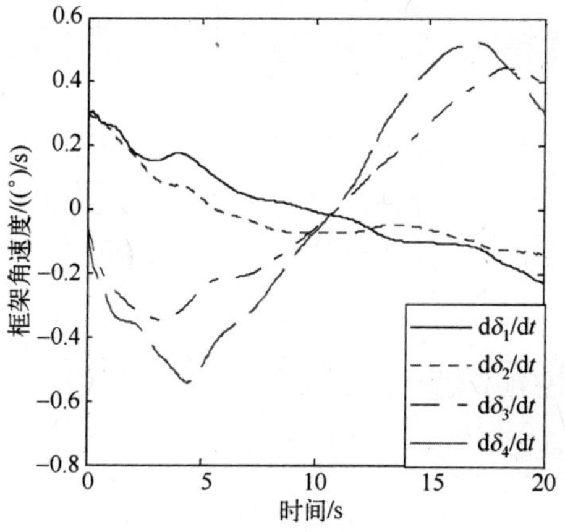

图 6.17　框架角速度变化曲线

由图 6.14 和图 6.15 可以看出，在有初始偏差和扰动的情况下，开环控制的精度是不够的，基于 LQR 的反馈控制能使系统轨迹收敛到最优姿态机动轨迹上，以实现整个系统的闭环操纵。图 6.16 和图 6.17 给出了 SGCMG 系统的框架角和角速度变化曲线，表明闭环控制量曲线是比较平滑的。由图 6.18 可以看出，相比开环控制，闭环控制的控制量发生一定变化，所以奇异度量值也有一定变化，但事先规划的最优姿态机动轨迹远离奇异状态，所以闭环控制结果也远离奇异状态。

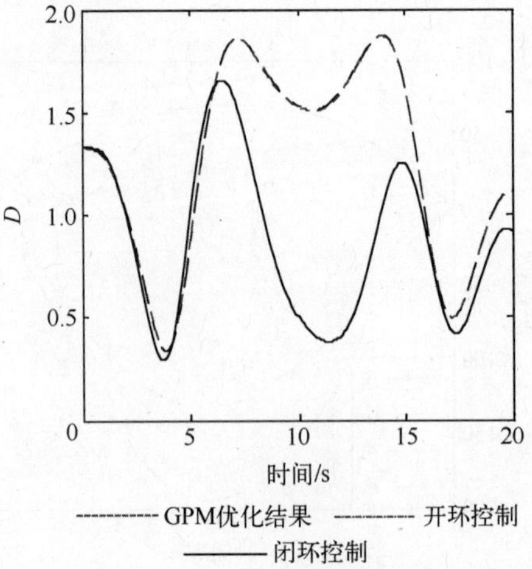

图 6.18　奇异量度变化曲线

6.2　空间高精度姿态机动的模型误差预测控制方法

针对参数不确定性以及外界干扰等情形，本节将研究模型误差预测控制方法（MEPC）。MEPC 是一种鲁棒控制方法，该方法包括一个标称控制器和一个模型误差估计器，利用预测滤波方法来确定模型误差，从而可以有效地补偿由参数模型误差和外界扰动引起的误差。

6.2.1　非线性预测滤波

考虑如下具有模型误差的非线性系统[9]：

$$\dot{\boldsymbol{x}}(t)=\boldsymbol{f}(\boldsymbol{x}(t))+\boldsymbol{G}(\boldsymbol{x}(t))\boldsymbol{d}(t) \tag{6.32}$$

$$\hat{\boldsymbol{y}}(t)=\boldsymbol{h}(\hat{\boldsymbol{x}}(t)) \tag{6.33}$$

式中：$\boldsymbol{x}$ 为 n 维的状态矢量；$\boldsymbol{f}$ 为 n 维非线性矢量；$\boldsymbol{d}$ 为由所有修正项构成的 l 维未知模型误差矢量；$\boldsymbol{G}$ 为相应的模型误差分布矩阵；$\hat{\boldsymbol{y}}$为 m 维的观测矢量；$\boldsymbol{v}$ 为 m 维的测量白噪声矢量。

对方程（6.33）中的$\hat{\boldsymbol{y}}(t)$进行泰勒级数展开，可得

$$\hat{\boldsymbol{y}}(t+\Delta t)=\hat{\boldsymbol{y}}(t)+\boldsymbol{z}[\hat{\boldsymbol{x}}(t),\Delta t]+\boldsymbol{\Lambda}(\Delta t)\boldsymbol{S}[\hat{\boldsymbol{x}}(t)]\boldsymbol{d}(t) \tag{6.34}$$

式中：$\Delta t=t_{k+1}-t_k$ 是采样时间间隔；$\boldsymbol{z}[\hat{\boldsymbol{x}}(t),\Delta t]$的第 i 个元素，计算公式为

$$z_i[\hat{\boldsymbol{x}}(t),\Delta t]=\sum_{k=1}^{p_i}\frac{\Delta t^k}{k!}L_f^k(h_i),\quad i=1,2,\cdots,m \tag{6.35}$$

其中，p_i 是对 $h_i[\hat{\boldsymbol{x}}(t),t]$进行连续微分，并将$\dot{\boldsymbol{x}}(t)$代入后，$\boldsymbol{d}(t)$的任一元素

首次出现时 $h_i[\hat{\boldsymbol{x}}(t),t]$ 的最低微分阶次，p_i 又称为相对阶；$L_f^k(h_i)$ 是 h_i 关于 $\boldsymbol{f}$ 的 k 阶李导数，定义为

$$\begin{cases} L_f^k(h_i)=h_i, & k=0 \\ L_f^k(h_i)=\dfrac{\partial L_f^{k-1}(h_i)}{\partial \hat{\boldsymbol{x}}}\boldsymbol{f}, & k\geqslant 1 \end{cases} \tag{6.36}$$

$\boldsymbol{\Lambda}(\Delta t)$ 是 $m\times m$ 维的对角阵，其对角线上的元素为

$$\lambda_{ii}=\frac{\Delta t^{p_i}}{p_i!},\quad i=1,2,\cdots,m \tag{6.37}$$

$\boldsymbol{S}[\hat{\boldsymbol{x}}(t)]$ 是 $m\times l$ 维的矩阵，其行元素为

$$s_i=\{L_{g_1}[L_f^{p_i-1}(h_i)],\cdots,L_{g_l}[L_f^{p_i-1}(h_i)]\},\quad i=1,2,\cdots,m \tag{6.38}$$

式中：关于 $\boldsymbol{G}$ 的列矢量 g_j 的李导数定义为

$$L_{g_j}[L_f^{p_i-1}(h_i)]=\frac{\partial L_f^{p_i-1}}{\partial \hat{\boldsymbol{x}}}g_j,\quad j=1,2,\cdots,l \tag{6.39}$$

基于最小模型误差（Minimum Model Error，MME）估计准则，对非线性系统、选取如下的性能指标函数：

$$J[\boldsymbol{d}(t)]=\frac{1}{2}[\tilde{\boldsymbol{y}}(t+\Delta t)-\hat{\boldsymbol{y}}(t+\Delta t)]^{\mathrm{T}}\boldsymbol{R}^{-1}[\tilde{\boldsymbol{y}}(t+\Delta t)-\hat{\boldsymbol{y}}(t+\Delta t)]+\frac{1}{2}\boldsymbol{d}^{\mathrm{T}}(t)\boldsymbol{W}\boldsymbol{d}(t) \tag{6.40}$$

式中：$\boldsymbol{R}$ 为测量噪声协方差阵。

把展开式（6.34）代入式（6.40），并使 $J[\boldsymbol{d}(t)]$ 关于 $\boldsymbol{d}(t)$ 达到最小，可求得

$$\boldsymbol{W}\boldsymbol{d}(t)=[\boldsymbol{\Lambda}(\Delta t)\boldsymbol{S}(\hat{\boldsymbol{x}})]^{\mathrm{T}}\boldsymbol{R}^{-1}[\tilde{\boldsymbol{y}}(t+\Delta t)-\hat{\boldsymbol{y}}(t+\Delta t)] \tag{6.41}$$

从而求得满足协方差约束条件的最小模型误差解：

$$\begin{aligned}\boldsymbol{d}(t)=-\{&[\boldsymbol{\Lambda}(\Delta t)\boldsymbol{S}(\hat{\boldsymbol{x}})]^{\mathrm{T}}\boldsymbol{R}^{-1}\boldsymbol{\Lambda}(\Delta t)\boldsymbol{S}(\hat{\boldsymbol{x}})+\boldsymbol{W}\}^{-1}\\ &[\boldsymbol{\Lambda}(\Delta t)\boldsymbol{S}(\hat{\boldsymbol{x}})]^{\mathrm{T}}\boldsymbol{R}^{-1}[\boldsymbol{z}(\hat{\boldsymbol{x}},\Delta t)+\hat{\boldsymbol{y}}(t)-\tilde{\boldsymbol{y}}(t+\Delta t)]\end{aligned} \tag{6.42}$$

非线性预测滤波器的基本原理：设已获得 t_k 时刻的状态估值，在接到 $t_k+\Delta t$ 时刻的测量信息 $\boldsymbol{y}(t_{k+1})$ 后，利用式（6.42）可预测 $[t_k,t_{k+1}]$ 间隔内的模型误差校正项 $\boldsymbol{d}(t)$，然后把 $\boldsymbol{d}(t)$ 的预测值代入式（6.32）中，将状态估值传播到 t_{k+1} 时刻。

上述过程的实质是先预测下一时刻的模型误差校正量，然后再对状态估计值进行传播，一步预测滤波之称即由此而得。图 6.19 所示为其滤波流程。

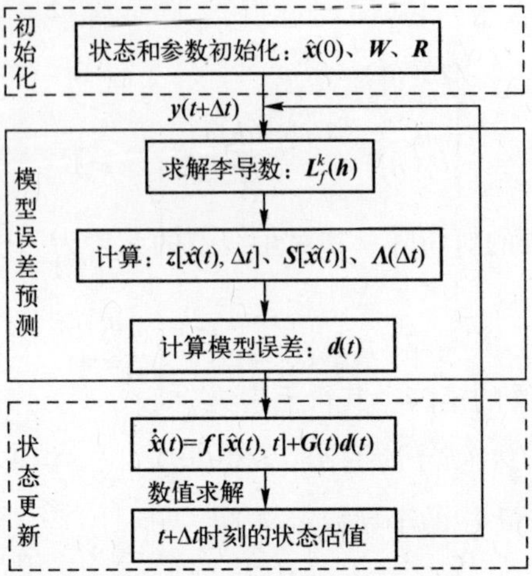

图 6.19　非线性预测滤波流程

6.2.2　模型误差预测控制方法

考虑非线性系统：

$$\dot{\boldsymbol{x}}(t)=\boldsymbol{f}(\boldsymbol{x}(t))+\boldsymbol{B}(\boldsymbol{x}(t))\boldsymbol{u}(t)+\boldsymbol{G}(\boldsymbol{x}(t))\boldsymbol{d}(t) \tag{6.43}$$

式中：$\boldsymbol{x}$ 为 n 维的状态矢量；$\boldsymbol{u}(t)$ 为控制输入；$\boldsymbol{d}(t)$ 为真实的模型误差矢量；$\boldsymbol{f}$ 为 n 维非线性矢量；$\boldsymbol{B}(\boldsymbol{x}(t))$ 为控制输入矩阵；$\boldsymbol{G}(\boldsymbol{x}(t))$ 为模型误差分配矩阵。简单地选取模型误差分布矩阵：

$$\boldsymbol{G}(\boldsymbol{x}(t))=\boldsymbol{B}(\boldsymbol{x}(t)) \tag{6.44}$$

因此，式（6.43）可以改写为

$$\dot{\boldsymbol{x}}(t)=\boldsymbol{f}(\boldsymbol{x}(t))+\boldsymbol{B}(\boldsymbol{x}(t))[\boldsymbol{u}(t)+\boldsymbol{d}(t)] \tag{6.45}$$

当采用模型预测控制算法时，可以通过估计模型误差来补偿真实的模型误差。方程（6.45）可以写为

$$\dot{\boldsymbol{x}}(t)=\boldsymbol{f}(\boldsymbol{x}(t))+\boldsymbol{B}(\boldsymbol{x}(t))[\overline{\boldsymbol{u}}(t)+\boldsymbol{d}(t)-\hat{\boldsymbol{d}}(t)] \tag{6.46}$$

式中：$\overline{\boldsymbol{u}}(t)$ 为标称控制输入。整个控制量可以表示为

$$\boldsymbol{u}(t)=\overline{\boldsymbol{u}}(t)-\hat{\boldsymbol{d}}(t) \tag{6.47}$$

系统的响应要先于模型误差的估计，因此计算控制力矩时需要用 $t-\Delta t$ 时刻估计的模型误差。控制输入可以表示为

$$\boldsymbol{u}(t)=\overline{\boldsymbol{u}}(t)-\hat{\boldsymbol{d}}(t-\Delta t) \tag{6.48}$$

根据第 2 章的内容，航天器姿态运动学和动力学方程可以描述为

$$\dot{\boldsymbol{q}}=\frac{1}{2}\boldsymbol{\Omega}(\boldsymbol{\omega})\boldsymbol{q}=\frac{1}{2}\boldsymbol{\Xi}(\boldsymbol{q})\boldsymbol{\omega} \tag{6.49}$$

$$\dot{\boldsymbol{\omega}}=-\boldsymbol{J}^{-1}[\boldsymbol{\omega}\times]\boldsymbol{J}\boldsymbol{\omega}+\boldsymbol{J}^{-1}\boldsymbol{T}+\boldsymbol{J}^{-1}\boldsymbol{d} \tag{6.50}$$

定义状态变量

$$\boldsymbol{x}\equiv\begin{bmatrix}\boldsymbol{x}_1\\ \boldsymbol{x}_2\end{bmatrix}=\begin{bmatrix}\boldsymbol{q}\\ \boldsymbol{\omega}\end{bmatrix} \tag{6.51}$$

为了进行全状态反馈，有

$$\boldsymbol{h}\equiv\begin{bmatrix}\boldsymbol{h}_1\\ \boldsymbol{h}_2\end{bmatrix}=\begin{bmatrix}\boldsymbol{q}\\ \boldsymbol{\omega}\end{bmatrix} \tag{6.52}$$

式（6.49）和式（6.50）可以写为如下状态方程形式：

$$\dot{\boldsymbol{x}}=\begin{bmatrix}\dot{\boldsymbol{q}}\\ \dot{\boldsymbol{\omega}}\end{bmatrix}=\begin{bmatrix}\frac{1}{2}\boldsymbol{\Xi}(\boldsymbol{q})\boldsymbol{\omega}\\ -\boldsymbol{J}^{-1}([\boldsymbol{\omega}\times]\boldsymbol{J}\boldsymbol{\omega}-\boldsymbol{T})\end{bmatrix}+\begin{bmatrix}\boldsymbol{0}_{4\times3}\\ \boldsymbol{J}^{-1}\end{bmatrix}\boldsymbol{d} \tag{6.53}$$

分析可知，$n=m=7$，从运动方程可以看到，控制输入 $\boldsymbol{T}$ 出现在 $\boldsymbol{\omega}$ 的一阶导数和 $\boldsymbol{q}$ 的二阶导数之中。因此，有

$$\boldsymbol{p}=[2\quad 2\quad 2\quad 2\quad 1\quad 1\quad 1]^{\mathrm{T}} \tag{6.54}$$

$$\boldsymbol{\Lambda}(\Delta t)=\begin{bmatrix}\frac{\Delta t^2}{2}\boldsymbol{I}_{4\times4} & \boldsymbol{0}_{4\times3}\\ \boldsymbol{0}_{3\times4} & \Delta t\boldsymbol{I}_{3\times3}\end{bmatrix} \tag{6.55}$$

为了描述 $\boldsymbol{z}$ 和 $\boldsymbol{S}$ 的表达式，需要先求出李导数，即

$$L_f^0(\boldsymbol{h}_1)=\boldsymbol{h}_1=\boldsymbol{q} \tag{6.56}$$

$$L_f^1(\boldsymbol{h}_1)=\frac{\partial L_f^0(\boldsymbol{h}_1)}{\partial \boldsymbol{x}}\boldsymbol{f}=\frac{\partial \boldsymbol{q}}{\partial \boldsymbol{x}}\boldsymbol{f} \tag{6.57}$$

四元数相对于状态变量的偏导数为

$$\frac{\partial \boldsymbol{q}}{\partial \boldsymbol{x}}=[\boldsymbol{I}_{4\times4}\quad \boldsymbol{0}_{4\times3}] \tag{6.58}$$

所以，式（6.57）变为

$$L_f^1(\boldsymbol{h}_1)=[\boldsymbol{I}_{4\times4}\quad \boldsymbol{0}_{4\times3}]\begin{bmatrix}\frac{1}{2}\boldsymbol{\Xi}(\boldsymbol{q})\boldsymbol{\omega}\\ -\boldsymbol{J}^{-1}([\boldsymbol{\omega}\times]\boldsymbol{J}\boldsymbol{\omega}-\boldsymbol{T})\end{bmatrix}=\frac{1}{2}\boldsymbol{\Xi}(\boldsymbol{q})\boldsymbol{\omega}=\frac{1}{2}\boldsymbol{\Omega}(\boldsymbol{\omega})\boldsymbol{q} \tag{6.59}$$

$\boldsymbol{h}_1$ 的二阶李导数为

$$L_f^2(\boldsymbol{h}_1)=\frac{\partial L_f^1(\boldsymbol{h}_1)}{\partial \boldsymbol{x}}\boldsymbol{f}$$

$$=\begin{bmatrix}\frac{1}{2}\boldsymbol{\Omega}(\boldsymbol{\omega}) & \frac{1}{2}\boldsymbol{\Xi}(\boldsymbol{q})\end{bmatrix}\begin{bmatrix}\frac{1}{2}\boldsymbol{\Xi}(\boldsymbol{q})\boldsymbol{\omega}\\ -\boldsymbol{J}^{-1}([\boldsymbol{\omega}\times]\boldsymbol{J}\boldsymbol{\omega}-\boldsymbol{T})\end{bmatrix} \tag{6.60}$$

$$=\frac{1}{4}\boldsymbol{\Omega}(\boldsymbol{\omega})\boldsymbol{\Omega}(\boldsymbol{\omega})\boldsymbol{q}-\frac{1}{2}\boldsymbol{\Xi}(\boldsymbol{q})\boldsymbol{J}^{-1}([\boldsymbol{\omega}\times]\boldsymbol{J}\boldsymbol{\omega}-\boldsymbol{T})$$

$\boldsymbol{h}_2$ 的李导数为

$$L_f^0(\boldsymbol{h}_2)=\boldsymbol{h}_2=\boldsymbol{\omega} \tag{6.61}$$

$$L_f^1(\boldsymbol{h}_2)=\frac{\partial L_f^0(\boldsymbol{h}_2)}{\partial \boldsymbol{x}}\boldsymbol{f}=\frac{\partial \boldsymbol{\omega}}{\partial \boldsymbol{x}}\boldsymbol{f}$$

$$=[\boldsymbol{0}_{3\times 4}\quad \boldsymbol{I}_{3\times 3}]\begin{bmatrix}\frac{1}{2}\boldsymbol{\Xi}(\boldsymbol{q})\boldsymbol{\omega}\\ -\boldsymbol{J}^{-1}([\boldsymbol{\omega}\times]\boldsymbol{J}\boldsymbol{\omega}-\boldsymbol{T})\end{bmatrix} \tag{6.62}$$

$$=-\boldsymbol{J}^{-1}([\boldsymbol{\omega}\times]\boldsymbol{J}\boldsymbol{\omega}-\boldsymbol{T})$$

剩下的李导数为

$$L_g[L_f^1(\boldsymbol{h}_1)]=\frac{\partial L_f^1(\boldsymbol{h}_1)}{\partial \boldsymbol{x}}\boldsymbol{G}$$

$$=\begin{bmatrix}\frac{1}{2}\boldsymbol{\Omega}(\boldsymbol{\omega}) & \frac{1}{2}\boldsymbol{\Xi}(\boldsymbol{q})\end{bmatrix}\begin{bmatrix}\boldsymbol{0}_{4\times 3}\\ \boldsymbol{J}^{-1}\end{bmatrix} \tag{6.63}$$

$$=\frac{1}{2}\boldsymbol{\Xi}(\boldsymbol{q})\boldsymbol{J}^{-1}$$

$$L_g[L_f^0(\boldsymbol{h}_2)]=\frac{\partial L_f^0(\boldsymbol{h}_2)}{\partial \boldsymbol{x}}\boldsymbol{G}$$

$$=[\boldsymbol{0}_{3\times 4}\quad \boldsymbol{I}_{3\times 3}]\begin{bmatrix}\boldsymbol{0}_{4\times 3}\\ \boldsymbol{J}^{-1}\end{bmatrix} \tag{6.64}$$

$$=\boldsymbol{J}^{-1}$$

利用上述求解的李导数，变量 z 和 $\boldsymbol{S}$ 计算如下：

$$z\equiv\begin{bmatrix}z_1\\ z_2\end{bmatrix} \tag{6.65}$$

其中，

$$
\begin{aligned}
z_1 &= \Delta t L_f^1(\boldsymbol{h}_1)+\frac{\Delta t^2}{2}L_f^2(\boldsymbol{h}_1)\\
&= \frac{\Delta t}{2}\boldsymbol{\Xi}(\boldsymbol{q})\boldsymbol{\omega}+\frac{\Delta t^2}{2}\left[\frac{1}{4}\boldsymbol{\Omega}(\boldsymbol{\omega})\boldsymbol{\Omega}(\boldsymbol{\omega})\boldsymbol{q}-\frac{1}{2}\boldsymbol{\Xi}(\boldsymbol{q})\boldsymbol{J}^{-1}([\boldsymbol{\omega}\times]\boldsymbol{J}\boldsymbol{\omega}-\boldsymbol{T})\right]
\end{aligned}
\tag{6.66}
$$

利用等式 $\boldsymbol{\Omega}(\boldsymbol{\omega})\boldsymbol{\Omega}(\boldsymbol{\omega})=-\boldsymbol{\omega}^{\mathrm{T}}\boldsymbol{\omega}\boldsymbol{I}_{4\times4}$，则式（6.66）可进一步改写为

$$
z_1=\frac{\Delta t}{2}\boldsymbol{\Xi}(\boldsymbol{q})\boldsymbol{\omega}-\frac{\Delta t^2}{4}\left[\frac{1}{2}(\boldsymbol{\omega}^{\mathrm{T}}\boldsymbol{\omega})\boldsymbol{q}+\boldsymbol{\Xi}(\boldsymbol{q})\boldsymbol{J}^{-1}([\boldsymbol{\omega}\times]\boldsymbol{J}\boldsymbol{\omega}-\boldsymbol{T})\right] \tag{6.67}
$$

矢量 z_2 为

$$
\begin{aligned}
z_2 &= \Delta t L_f^1(\boldsymbol{h}_2)\\
&= -\Delta t\boldsymbol{J}^{-1}([\boldsymbol{\omega}\times]\boldsymbol{J}\boldsymbol{\omega}-\boldsymbol{T})
\end{aligned}
\tag{6.68}
$$

矩阵 $\boldsymbol{S}$ 可以表示为

$$
\boldsymbol{S}=\begin{bmatrix}\dfrac{1}{2}\boldsymbol{\Xi}(\boldsymbol{q})\boldsymbol{J}^{-1}\\ \boldsymbol{J}^{-1}\end{bmatrix} \tag{6.69}
$$

利用这些定义，则可根据 6.2.1 节的内容估计模型误差 $\boldsymbol{d}(t)$。

6.2.3　姿态控制器设计

常见的姿态控制任务主要包括姿态调节、姿态机动和姿态跟踪三大类任务。从数学模型来看，姿态调节和姿态机动可看成姿态跟踪的特例。因此，本节以姿态跟踪任务为例，推导基于四元数的航天器姿态跟踪误差模型，并分别设计 PD、VSC 以及 LQR 三种姿态跟踪控制器。

6.2.3.1　航天器姿态跟踪误差模型

在航天器姿态跟踪任务中，常用单位四元数 $\boldsymbol{q}_t=[\boldsymbol{\varrho}_t^{\mathrm{T}}\quad q_{t4}]^{\mathrm{T}}$ 表示期望坐标系相对于惯性坐标系的姿态四元数，$\boldsymbol{\omega}_t$ 表示期望坐标系相对于惯性坐标系的角速度矢量。定义如下姿态误差四元数和角速度误差：

$$
\boldsymbol{q}_e=\boldsymbol{q}\otimes\boldsymbol{q}_t^{-1} \tag{6.70}
$$

$$
\boldsymbol{\omega}_e=\boldsymbol{\omega}-\delta\boldsymbol{A}\boldsymbol{\omega}_t \tag{6.71}
$$

式中：$\boldsymbol{q}$ 和 $\boldsymbol{\omega}$ 分别为航天器当前姿态和角速度；$\delta\boldsymbol{A}$ 表示航天器实际姿态与期望姿态之间的姿态误差阵，由误差四元数 $\boldsymbol{q}_e$ 表示为

$$
\delta\boldsymbol{A}(\boldsymbol{q}_e)=(q_{e4}^2-\|\boldsymbol{\varrho}_e\|^2)\boldsymbol{I}_{3\times3}+2\boldsymbol{\varrho}_e\boldsymbol{\varrho}_e^{\mathrm{T}}-2q_{e4}[\boldsymbol{\varrho}_e\times] \tag{6.72}
$$

根据式（2.46）、式（2.18）、式（6.70）和式（6.71）可得航天器跟踪误差动力学和运动学方程为

$$
\begin{aligned}
\boldsymbol{J}\dot{\boldsymbol{\omega}}_e &= -\boldsymbol{\omega}\times\boldsymbol{J}\boldsymbol{\omega}-\boldsymbol{J}(\delta\boldsymbol{A}\dot{\boldsymbol{\omega}}_t-\boldsymbol{\omega}_e\times\delta\boldsymbol{A}\boldsymbol{\omega}_t)+\boldsymbol{T}+\boldsymbol{d}\\
&= -(\boldsymbol{\omega}_e+\delta\boldsymbol{A}\boldsymbol{\omega}_t)\times\boldsymbol{J}(\boldsymbol{\omega}_e+\delta\boldsymbol{A}\boldsymbol{\omega}_t)-\boldsymbol{J}(\delta\boldsymbol{A}\dot{\boldsymbol{\omega}}_t-\boldsymbol{\omega}_e\times\delta\boldsymbol{A}\boldsymbol{\omega}_t)+\boldsymbol{T}+\boldsymbol{d}
\end{aligned}
\tag{6.73}
$$

$$\begin{cases}\dot{\boldsymbol{\varrho}}_e=\dfrac{1}{2}([\boldsymbol{\varrho}_e\times]+q_{e4}\boldsymbol{I}_{3\times3})\boldsymbol{\omega}_e\\ \dot{q}_{e4}=-\dfrac{1}{2}\boldsymbol{\varrho}_e^{\mathrm{T}}\boldsymbol{\omega}_e\end{cases}\tag{6.74}$$

6.2.3.2 PD 控制器

对于刚体航天器，若已知航天器的惯量张量矩阵，则可以实现绕欧拉轴机动，从而将三轴转动机动问题转化为单轴转动机动问题，进而实现最优或近似最优路径规划。设计 PD 控制律为

$$\boldsymbol{T}=-k_\omega\boldsymbol{\omega}_e-k_q\boldsymbol{\varrho}_e+\mathrm{RNR}\tag{6.75}$$

式中：RNR 为非线性转动项，即

$$\mathrm{RNR}=[\boldsymbol{\omega}\times]\boldsymbol{J}\boldsymbol{\omega}+\boldsymbol{J}(\delta\boldsymbol{A}\dot{\boldsymbol{\omega}}_t-\boldsymbol{\omega}_e\times\delta\boldsymbol{A}\boldsymbol{\omega}_t)\tag{6.76}$$

利用等式 $\delta\boldsymbol{A}\boldsymbol{\omega}_t\times\delta\boldsymbol{A}\boldsymbol{\omega}_t=\boldsymbol{0}$，则可进一步得到 PD 形式为

$$\boldsymbol{T}=[\boldsymbol{\omega}\times]\boldsymbol{J}\boldsymbol{\omega}+\boldsymbol{J}(\delta\boldsymbol{A}\dot{\boldsymbol{\omega}}_t-[\boldsymbol{\omega}\times]\delta\boldsymbol{A}\boldsymbol{\omega}_t)-k_\omega\boldsymbol{\omega}_e-k_q\boldsymbol{\varrho}_e\tag{6.77}$$

该形式中补偿的非线性转动项直接与航天器姿态角速度关联，该角速度可由陀螺测量得到，或者状态滤波器估计得到，相关原理见参考文献［10］。

如果进一步考虑干扰力矩的影响，可以构造如下姿态控制律：

$$\boldsymbol{T}=[\boldsymbol{\omega}\times]\boldsymbol{J}\boldsymbol{\omega}+\boldsymbol{J}(\delta\boldsymbol{A}\dot{\boldsymbol{\omega}}_t-[\boldsymbol{\omega}\times]\delta\boldsymbol{A}\boldsymbol{\omega}_t)-k_\omega\boldsymbol{\omega}_e-k_q\boldsymbol{\varrho}_e-d_{\max}\mathrm{sgn}(\boldsymbol{\omega}_e)\tag{6.78}$$

式中：$k_\omega>0$，$k_q>0$，$d_{\max}$为干扰力矩的最大边界值。

分析可知，对于由式（6.73）、式（6.74）和式（6.78）组成的闭环控制系统，$\boldsymbol{\omega}_e=\boldsymbol{0}$，$\boldsymbol{q}_e=[0\quad0\quad0\quad1]^{\mathrm{T}}$ 和 $\boldsymbol{\omega}_e=\boldsymbol{0}$，$\boldsymbol{q}_e=[0\quad0\quad0\quad-1]^{\mathrm{T}}$ 均是系统的平衡点，至于具体趋于哪种状态，与误差四元数标量分量的初始值符号相关。

下面对所设计的姿态控制律进行稳定性分析，构造状态矢量 $\boldsymbol{x}=[\boldsymbol{\omega}_e^{\mathrm{T}}\quad\boldsymbol{q}_e^{\mathrm{T}}]^{\mathrm{T}}$，并选择李雅普诺夫函数

$$\begin{aligned}V(\boldsymbol{x})&=\frac{1}{2}\boldsymbol{\omega}_e^{\mathrm{T}}\boldsymbol{J}\boldsymbol{\omega}_e+k_q\boldsymbol{\varrho}_e^{\mathrm{T}}\boldsymbol{\varrho}_e+k_q(1-q_{e4})^2\\&=\frac{1}{2}\boldsymbol{\omega}_e^{\mathrm{T}}\boldsymbol{J}\boldsymbol{\omega}_e+2k_q(1-q_{e4})\end{aligned}\tag{6.79}$$

显然，$V(\boldsymbol{x})$正定，且关于 $\boldsymbol{x}$ 径向无界，即当$\|\boldsymbol{x}\|\to\infty$时，有 $V(\boldsymbol{x})\to\infty$。

对式（6.79）进行求导可得

$$\begin{aligned}\dot{V}(\boldsymbol{x})&=\boldsymbol{\omega}_e^{\mathrm{T}}\boldsymbol{J}\dot{\boldsymbol{\omega}}_e-2k_q\dot{q}_{e4}\\&=\boldsymbol{\omega}_e^{\mathrm{T}}\{-[\boldsymbol{\omega}\times]\boldsymbol{J}\boldsymbol{\omega}-\boldsymbol{J}(\delta\boldsymbol{A}\dot{\boldsymbol{\omega}}_t-\boldsymbol{\omega}_e\times\delta\boldsymbol{A}\boldsymbol{\omega}_t)+\boldsymbol{T}+\boldsymbol{d}\}+k_q\boldsymbol{\varrho}_e^{\mathrm{T}}\boldsymbol{\omega}_e\end{aligned}\tag{6.80}$$

将式（6.78）中的控制律 $\boldsymbol{T}$ 代入式（6.80），可得

$$\begin{aligned}\dot{V}(\boldsymbol{x}) &= -k_{\omega}\boldsymbol{\omega}_e^{\mathrm{T}}\boldsymbol{\omega}_e - k_q\boldsymbol{\omega}_e^{\mathrm{T}}\boldsymbol{\varrho}_e - d_{\max}\boldsymbol{\omega}_e^{\mathrm{T}}\mathrm{sgn}(\boldsymbol{\omega}_e) + \boldsymbol{\omega}_e^{\mathrm{T}}\boldsymbol{d} + k_q\boldsymbol{\varrho}_e^{\mathrm{T}}\boldsymbol{\omega}_e \\ &= -k_{\omega}\boldsymbol{\omega}_e^{\mathrm{T}}\boldsymbol{\omega}_e - d_{\max}\boldsymbol{\omega}_e^{\mathrm{T}}\mathrm{sgn}(\boldsymbol{\omega}_e) + \boldsymbol{\omega}_e^{\mathrm{T}}\boldsymbol{d} \leqslant 0\end{aligned} \tag{6.81}$$

这表明 $\dot{V}(\boldsymbol{x})$ 是负半定的。记 $\mathcal{M} = \{\boldsymbol{x} \mid \dot{V}(\boldsymbol{x}) = 0\}$ 是系统的最大不变集，由式（6.81）可知 $\dot{V}(\boldsymbol{x}) = 0 \Rightarrow \boldsymbol{\omega}_e = \boldsymbol{0}$，则有 $\mathcal{M} = \{\boldsymbol{x} \mid \boldsymbol{\omega}_e = \boldsymbol{0}\}$。设 $\boldsymbol{x}(t)$ 是属于 $\mathcal{M}$ 的一个解，由 $\boldsymbol{\omega}_e(t) \equiv \boldsymbol{0}$ 可得 $\dot{\boldsymbol{\omega}}_e(t) \equiv \boldsymbol{0}$，再利用式（6.73）和式（6.78）可得 $\boldsymbol{\varrho}_e(t) \equiv \boldsymbol{0}$，将其代入式（6.74）有 $q_{e4} = \pm 1$。由 LaSalle 不变集定理可知，平衡点是渐近稳定的，由于所选定的 Lyapunov 函数是径向无界的，所以闭环系统是全局渐近稳定的，证毕。

进一步，考虑姿态角速度受限和输入受限情形时，可选用如下的 PID 控制器[11]：

$$\begin{aligned}\boldsymbol{T} &= -\boldsymbol{K}_q \underset{L_i}{\mathrm{sat}}\left(\boldsymbol{\varrho}_e + \frac{1}{T}\int \boldsymbol{\varrho}_e\right) - \boldsymbol{K}_{\omega}\boldsymbol{\omega}_e \\ &= -\boldsymbol{J}\left\{2k_q \underset{L_i}{\mathrm{sat}}\left(\boldsymbol{\varrho}_e + \frac{1}{T}\int \boldsymbol{\varrho}_e\right) + k_{\omega}\boldsymbol{\omega}_e\right\}\end{aligned} \tag{6.82}$$

式中：系数矩阵 $\boldsymbol{K}_q = 2k_q\boldsymbol{J}$，$\boldsymbol{K}_{\omega} = k_{\omega}\boldsymbol{J}$。输入受限 L_i 为

$$L_i = \frac{k_{\omega}}{2k_q}\min\left\{\sqrt{4a_i \mid q_{ei} \mid},\ \mid \omega_i \mid_{\max}\right\} \tag{6.83}$$

$$-U \leqslant u_i(t) \leqslant U \tag{6.84}$$

式中：U 为控制力矩限幅；$a_i = U/J_{ii}$；$|\omega_i|_{\max}$ 为给定的第 i 轴姿态角速度限幅。控制器增益可以确定为 $k_q = \omega_n^2 + 2\zeta\omega_n/T$，$k_{\omega} = 2\zeta\omega_n + 1/T$，$\omega_n$ 为自然频率，ζ 为阻尼比，积分控制时间常数 T 一般选取为 $T \approx 10/(\zeta\omega_n)$。式（6.82）中的饱和函数定义为

$$\underset{L}{\mathrm{sat}}(x) = \begin{cases} x, & \|x\|_{\infty} < L \\ L\dfrac{x}{\|x\|_{\infty}}, & \|x\|_{\infty} \geqslant L \end{cases} \tag{6.85}$$

6.2.3.3 VSC 控制器

考虑具有一般形式的动态系统

$$\dot{\boldsymbol{x}} = \boldsymbol{f}(\boldsymbol{x}) + \boldsymbol{B}(\boldsymbol{x})\boldsymbol{u} \tag{6.86}$$

设计的滑模控制律通常可表示为

$$\boldsymbol{u}_{\mathrm{vsc}} = -\boldsymbol{B}^{-1}\left[\boldsymbol{f} + \dot{\boldsymbol{s}} - \boldsymbol{x}^{(n)} + \eta\,\mathrm{sat}\left(\frac{\boldsymbol{s}}{\varepsilon}\right)\right] \tag{6.87}$$

式中：$\boldsymbol{s}$ 为滑动矢量，也称为滑模面；η 和 ε（边界层厚度）为额外的控制增益，饱和函数 sat 用于减弱抖振。

若不考虑干扰力矩，则刚体航天器姿态动力学方程为

$$\dot{\boldsymbol{\omega}}=-\boldsymbol{J}^{-1}[\boldsymbol{\omega}\times]\boldsymbol{J}\boldsymbol{\omega}+\boldsymbol{J}^{-1}\boldsymbol{T} \tag{6.88}$$

选择滑动矢量

$$\boldsymbol{s}=(\boldsymbol{\omega}-\boldsymbol{\omega}_t)+\lambda\cdot\operatorname{sgn}(q_{e4})\boldsymbol{\Xi}^{\mathrm{T}}(\boldsymbol{q}_t)\boldsymbol{q} \tag{6.89}$$

滑模面的时间微分为

$$\dot{\boldsymbol{s}}=(\dot{\boldsymbol{\omega}}-\dot{\boldsymbol{\omega}}_t)+\lambda\cdot\operatorname{sgn}(q_{e4})(\boldsymbol{\Xi}^{\mathrm{T}}(\dot{\boldsymbol{q}}_t)\boldsymbol{q}+\boldsymbol{\Xi}^{\mathrm{T}}(\boldsymbol{q}_t)\dot{\boldsymbol{q}}) \tag{6.90}$$

利用等式

$$\boldsymbol{\Xi}^{\mathrm{T}}(\boldsymbol{a})\boldsymbol{b}=-\boldsymbol{\Xi}^{\mathrm{T}}(\boldsymbol{b})\boldsymbol{a} \tag{6.91}$$

进一步，式（6.90）可以改写为

$$\begin{aligned}\dot{\boldsymbol{s}}&=(\dot{\boldsymbol{\omega}}-\dot{\boldsymbol{\omega}}_t)+\lambda\cdot\operatorname{sgn}(q_{e4})(-\boldsymbol{\Xi}^{\mathrm{T}}(\boldsymbol{q})\dot{\boldsymbol{q}}_t+\boldsymbol{\Xi}^{\mathrm{T}}(\boldsymbol{q}_t)\dot{\boldsymbol{q}})\\&=(\dot{\boldsymbol{\omega}}-\dot{\boldsymbol{\omega}}_t)+\frac{\lambda}{2}\cdot\operatorname{sgn}(q_{e4})(-\boldsymbol{\Xi}^{\mathrm{T}}(\boldsymbol{q})\boldsymbol{\Xi}(\boldsymbol{q}_t)\boldsymbol{\omega}_t+\boldsymbol{\Xi}^{\mathrm{T}}(\boldsymbol{q}_t)\boldsymbol{\Xi}(\boldsymbol{q})\boldsymbol{\omega})\end{aligned} \tag{6.92}$$

由式（6.88）可知

$$\boldsymbol{f}=-\boldsymbol{J}^{-1}[\boldsymbol{\omega}\times]\boldsymbol{J}\boldsymbol{\omega} \tag{6.93}$$

$$\boldsymbol{B}=\boldsymbol{J}^{-1} \tag{6.94}$$

将式（6.89）和式（6.92）~式（6.94）代入式（6.87）可得

$$\boldsymbol{T}=[\boldsymbol{\omega}\times]\boldsymbol{J}\boldsymbol{\omega}+\boldsymbol{J}\left\{\frac{\lambda}{2}\cdot\operatorname{sgn}(q_{e4})[\boldsymbol{\Xi}^{\mathrm{T}}(\boldsymbol{q})\boldsymbol{\Xi}(\boldsymbol{q}_t)\boldsymbol{\omega}_t-\boldsymbol{\Xi}^{\mathrm{T}}(\boldsymbol{q}_t)\boldsymbol{\Xi}(\boldsymbol{q})\boldsymbol{\omega}]+\dot{\boldsymbol{\omega}}_t-\boldsymbol{G}\boldsymbol{\vartheta}\right\} \tag{6.95}$$

式中：$\boldsymbol{G}$ 是一个正定的 3×3 对角矩阵，$\boldsymbol{\vartheta}$ 的第 i 个分量为

$$\vartheta_i=\operatorname{sat}(s_i,\varepsilon_i) \tag{6.96}$$

其中，

$$\operatorname{sat}(s_i,\varepsilon_i)\equiv\begin{cases}1, & s_i>\varepsilon_i\\ s_i/\varepsilon_i, & |s_i|\leqslant\varepsilon_i\\ -1 & s_i<-\varepsilon_i\end{cases} \tag{6.97}$$

另外，式（6.89）中的滑动矢量也可以选为

$$\boldsymbol{s}=(\boldsymbol{\omega}-\delta\boldsymbol{A}\boldsymbol{\omega}_t)+\lambda\cdot\operatorname{sgn}(q_{e4})\boldsymbol{\Xi}^{\mathrm{T}}(\boldsymbol{q}_t)\boldsymbol{q} \tag{6.98}$$

式中：$\delta\boldsymbol{A}$ 为从期望姿态到实际姿态的姿态误差阵，则对应的控制律为

$$\boldsymbol{T}=[\boldsymbol{\omega}\times]\boldsymbol{J}\boldsymbol{\omega}+\boldsymbol{J}\left\{\frac{\lambda}{2}\cdot\operatorname{sgn}(q_{e4})[\boldsymbol{\Xi}^{\mathrm{T}}(\boldsymbol{q})\boldsymbol{\Xi}(\boldsymbol{q}_t)\boldsymbol{\omega}_t-\boldsymbol{\Xi}^{\mathrm{T}}(\boldsymbol{q}_t)\boldsymbol{\Xi}(\boldsymbol{q})\boldsymbol{\omega}]+\delta\boldsymbol{A}\dot{\boldsymbol{\omega}}_t-[\boldsymbol{\omega}\times]\delta\boldsymbol{A}\boldsymbol{\omega}_t-\boldsymbol{G}\boldsymbol{\vartheta}\right\} \tag{6.99}$$

下面对所设计的姿态控制律进行稳定性分析，选择 Lyapunov 函数

$$V=\frac{1}{2}\boldsymbol{s}^{\mathrm{T}}\boldsymbol{s} \tag{6.100}$$

对式（6.100）进行求导，并利用式（6.88）、式（6.98）和式（6.99）可得

$$\begin{aligned}\dot{V}&=\boldsymbol{s}^{\mathrm{T}}\dot{\boldsymbol{s}}\\&=-\boldsymbol{s}^{\mathrm{T}}\boldsymbol{G}\boldsymbol{\vartheta}\end{aligned} \tag{6.101}$$

由于 $\boldsymbol{G}$ 是正定矩阵，所以 $\dot{V}<0$，系统是全局渐近稳定的。

6.2.3.4　LQR 控制器

根据第 2 章的内容，航天器姿态四元数运动学和动力学方程分别为

$$\dot{\boldsymbol{q}}=\frac{1}{2}\boldsymbol{\Xi}(\boldsymbol{q})\boldsymbol{\omega}=\frac{1}{2}\boldsymbol{\Omega}(\boldsymbol{\omega})\boldsymbol{q} \tag{6.102}$$

$$\dot{\boldsymbol{\omega}}=-\boldsymbol{J}^{-1}[\boldsymbol{\omega}\times]\boldsymbol{J}\boldsymbol{\omega}+\boldsymbol{J}^{-1}\boldsymbol{T} \tag{6.103}$$

对于期望姿态四元数 $\boldsymbol{q}_t$，则有

$$\dot{\boldsymbol{q}}_t=\frac{1}{2}\boldsymbol{\Xi}(\boldsymbol{q}_t)\boldsymbol{\omega}_t \tag{6.104}$$

根据四元数乘法规则，式（6.70）中的误差四元数的矢量和标量部分又可分别表示为

$$\boldsymbol{\varrho}_e=\boldsymbol{\Xi}^{\mathrm{T}}(\boldsymbol{q}_t)\boldsymbol{q} \tag{6.105}$$

$$q_{e4}=\boldsymbol{q}_t^{\mathrm{T}}\boldsymbol{q} \tag{6.106}$$

注意到随着 $\boldsymbol{\varrho}_e$ 越接近零，实际的姿态四元数就越接近期望姿态四元数。设计二阶闭环系统

$$\ddot{\boldsymbol{\varrho}}_e+\boldsymbol{L}_2\dot{\boldsymbol{\varrho}}_e+\boldsymbol{L}_1\boldsymbol{\varrho}_e=\boldsymbol{0} \tag{6.107}$$

式中：$\boldsymbol{L}_1$ 和 $\boldsymbol{L}_2$ 分别是 3×3 的增益矩阵。这些矩阵可由 LQR 方法进行确定，令

$$\ddot{\boldsymbol{\varrho}}_e=\boldsymbol{u} \tag{6.108}$$

$$\boldsymbol{u}=-\boldsymbol{L}\begin{bmatrix}\boldsymbol{\varrho}_e\\\dot{\boldsymbol{\varrho}}_e\end{bmatrix} \tag{6.109}$$

式中：$\boldsymbol{L}\equiv[\boldsymbol{L}_1\quad\boldsymbol{L}_2]$。选取状态矢量 $\boldsymbol{x}\equiv[\boldsymbol{\varrho}_e^{\mathrm{T}}\quad\dot{\boldsymbol{\varrho}}_e^{\mathrm{T}}]^{\mathrm{T}}$，于是有

$$\dot{\boldsymbol{x}}=\begin{bmatrix}\boldsymbol{0}_{3\times3}&\boldsymbol{I}_{3\times3}\\\boldsymbol{0}_{3\times3}&\boldsymbol{0}_{3\times3}\end{bmatrix}\boldsymbol{x}+\begin{bmatrix}\boldsymbol{0}_{3\times3}\\\boldsymbol{I}_{3\times3}\end{bmatrix}\boldsymbol{u} \tag{6.110}$$

值得注意的是，若 $\boldsymbol{L}_1$ 和 $\boldsymbol{L}_2$ 是标量，则可针对式（6.107）利用二阶闭环系统性能设计参数，无须求解 LQR 问题。实际上，结合式（6.109）和式（6.110），矩阵 $\boldsymbol{L}$ 可看成式（6.110）中的状态方程的反馈增益矩阵。因

此，如果 $\boldsymbol{L}_1$ 和 $\boldsymbol{L}_2$ 是矩阵，可通过求解式（6.110）的 LQR 问题得到。进一步，由于式（6.110）是线性定常系统，在给定权重矩阵 $\boldsymbol{Q}$ 和 $\boldsymbol{R}$ 的情形下，求解的增益矩阵 $\boldsymbol{L}_1$ 和 $\boldsymbol{L}_2$ 也是常值矩阵，有利于进行在线计算。

对式（6.105）分别求一阶和二阶时间导数，并将它们代入式（6.107）可得

$$\boldsymbol{\Xi}^{\mathrm{T}}(\boldsymbol{q}_t)\ddot{\boldsymbol{q}}+[2\boldsymbol{\Xi}^{\mathrm{T}}(\dot{\boldsymbol{q}}_t)+\boldsymbol{L}_2\boldsymbol{\Xi}^{\mathrm{T}}(\boldsymbol{q}_t)]\dot{\boldsymbol{q}}+[\boldsymbol{\Xi}^{\mathrm{T}}(\ddot{\boldsymbol{q}}_t)+\boldsymbol{L}_2\boldsymbol{\Xi}^{\mathrm{T}}(\dot{\boldsymbol{q}}_t)+\boldsymbol{L}_1\boldsymbol{\Xi}^{\mathrm{T}}(\boldsymbol{q}_t)]\boldsymbol{q}=\boldsymbol{0} \tag{6.111}$$

对式（6.102）求导并利用等式 $\boldsymbol{\Omega}^2(\boldsymbol{\omega})=-(\boldsymbol{\omega}^{\mathrm{T}}\boldsymbol{\omega})\boldsymbol{I}_{4\times4}$，可得

$$\begin{aligned}\ddot{\boldsymbol{q}}&=\frac{1}{2}\boldsymbol{\Xi}(\boldsymbol{q})\dot{\boldsymbol{\omega}}+\frac{1}{2}\boldsymbol{\Omega}(\boldsymbol{\omega})\dot{\boldsymbol{q}}\\&=\frac{1}{2}\boldsymbol{\Xi}(\boldsymbol{q})\dot{\boldsymbol{\omega}}-\frac{1}{4}(\boldsymbol{\omega}^{\mathrm{T}}\boldsymbol{\omega})\boldsymbol{q}\end{aligned} \tag{6.112}$$

类似地，对于期望姿态四元数 $\boldsymbol{q}_t$，同样有

$$\ddot{\boldsymbol{q}}_t=\frac{1}{2}\boldsymbol{\Xi}(\boldsymbol{q}_t)\dot{\boldsymbol{\omega}}_t-\frac{1}{4}(\boldsymbol{\omega}_t^{\mathrm{T}}\boldsymbol{\omega}_t)\boldsymbol{q}_t \tag{6.113}$$

其中，$\dot{\boldsymbol{\omega}}_t$ 可由期望的动力学方程推导得到。将式（6.103）代入式（6.112）可得

$$\ddot{\boldsymbol{q}}=-\frac{1}{2}\boldsymbol{\Xi}(\boldsymbol{q})\boldsymbol{J}^{-1}[\boldsymbol{\omega}\times]\boldsymbol{J}\boldsymbol{\omega}-\frac{1}{4}(\boldsymbol{\omega}^{\mathrm{T}}\boldsymbol{\omega})\boldsymbol{q}+\frac{1}{2}\boldsymbol{\Xi}(\boldsymbol{q})\boldsymbol{J}^{-1}\boldsymbol{T} \tag{6.114}$$

将式（6.102）和式（6.114）代入式（6.111），求解控制力矩 $\boldsymbol{T}$ 可得

$$\begin{aligned}\boldsymbol{T}=&[\boldsymbol{\omega}\times]\boldsymbol{J}\boldsymbol{\omega}+2\boldsymbol{J}[\boldsymbol{\Xi}^{\mathrm{T}}(\boldsymbol{q}_t)\boldsymbol{\Xi}(\boldsymbol{q})]^{-1}\left\{\frac{1}{4}(\boldsymbol{\omega}^{\mathrm{T}}\boldsymbol{\omega})\boldsymbol{\Xi}^{\mathrm{T}}(\boldsymbol{q}_t)-\boldsymbol{\Xi}^{\mathrm{T}}(\dot{\boldsymbol{q}}_t)\boldsymbol{\Omega}(\boldsymbol{\omega})\right.\\&\left.-\boldsymbol{\Xi}^{\mathrm{T}}(\ddot{\boldsymbol{q}}_t)-\boldsymbol{L}_1\boldsymbol{\Xi}^{\mathrm{T}}(\boldsymbol{q}_t)-\boldsymbol{L}_2\left[\frac{1}{2}\boldsymbol{\Xi}^{\mathrm{T}}(\boldsymbol{q}_t)\boldsymbol{\Omega}(\boldsymbol{\omega})+\boldsymbol{\Xi}^{\mathrm{T}}(\dot{\boldsymbol{q}}_t)\right]\right\}\boldsymbol{q}\end{aligned} \tag{6.115}$$

式中：$\dot{\boldsymbol{q}}_t$ 和 $\ddot{\boldsymbol{q}}_t$ 分别可由式（6.104）和式（6.113）计算得到。分析可知，只要 $q_{e4}=\boldsymbol{q}_t^{\mathrm{T}}\boldsymbol{q}$ 非零，$\boldsymbol{\Xi}^{\mathrm{T}}(\boldsymbol{q}_t)\boldsymbol{\Xi}(\boldsymbol{q})$ 的逆总是存在。这可由下述等式得到

$$\boldsymbol{\Xi}^{\mathrm{T}}(\boldsymbol{q}_t)\boldsymbol{\Xi}(\boldsymbol{q})=q_{e4}\boldsymbol{I}_{3\times3}+[\boldsymbol{\varrho}_e\times] \tag{6.116}$$

$$[\boldsymbol{\Xi}^{\mathrm{T}}(\boldsymbol{q}_t)\boldsymbol{\Xi}(\boldsymbol{q})]^{-1}=q_{e4}\boldsymbol{I}_{3\times3}-[\boldsymbol{\varrho}_e\times]+\frac{\boldsymbol{\varrho}_e\boldsymbol{\varrho}_e^{\mathrm{T}}}{q_{e4}} \tag{6.117}$$

若 $\boldsymbol{L}_1$ 和 $\boldsymbol{L}_2$ 是标量，即 $\boldsymbol{L}_1=l_1$ 和 $\boldsymbol{L}_2=l_2$，则式（6.115）可进一步简化为

$$\boldsymbol{T}=[\boldsymbol{\omega}\times]\boldsymbol{J}\boldsymbol{\omega}+\boldsymbol{J}\left\{\delta\boldsymbol{A}\dot{\boldsymbol{\omega}}_t-[\boldsymbol{\omega}\times]\delta\boldsymbol{A}\boldsymbol{\omega}_t-l_2\boldsymbol{\omega}_e-2\left[\frac{4l_1-(\boldsymbol{\omega}_e^{\mathrm{T}}\boldsymbol{\omega}_e)}{4q_4}\right]\boldsymbol{\varrho}_e\right\} \tag{6.118}$$

式中：

$$\delta A = A(q)A^{\mathrm{T}}(q_t) \tag{6.119}$$

$$\omega_e = \omega - \delta A \omega_t \tag{6.120}$$

可以看出，式（6.118）中并不显含 $\dot{q}_t$ 和 $\ddot{q}_t$。

6.2.4　仿真实验与分析

本节针对存在较大干扰情况下的航天器姿态机动任务，除了采用滑模变结构控制（VSC）方法设计姿态控制律，还采用前述的模型误差预测控制算法（MEPC）补偿由参数模型误差和外界扰动引起的误差，数学仿真验证了其有效性。

在这里，实际惯量张量矩阵 $J=\bar{J}+\Delta J$，标称惯量张量阵 $\bar{J}$ 和惯量张量不确定部分 ΔJ 分别为

$$\bar{J} = \begin{bmatrix} 30 & 10 & 5 \\ 10 & 20 & 3 \\ 5 & 3 & 15 \end{bmatrix} \mathrm{kg \cdot m^2}, \quad \Delta J = \begin{bmatrix} 3 & 0 & 0 \\ 0 & 2 & 0 \\ 0 & 0 & 1.5 \end{bmatrix} \mathrm{kg \cdot m^2} \tag{6.121}$$

外部干扰力矩为

$$d = 1\times 10^{-2} \begin{bmatrix} \sin(10^{-4}t) \\ 1 \\ \cos(10^{-4}t) \end{bmatrix} \mathrm{N \cdot m} \tag{6.122}$$

本节采用 SGCMG 系统作为执行机构，相关数学模型见 2.4.2 节；并利用星敏感器对航天器进行姿态测量，测量精度为 10″，频率为 1Hz，相关原理详见参考文献［11］，这里不再赘述。

卫星初始姿态角速度和期望姿态角速度均为 $[0,0,0]$°/s；初始欧拉角 $[\varphi_0,\theta_0,\psi_0]$ 为 $[0°,0°,0°]$，期望欧拉角为 $[40°,30°,20°]$，仿真时间 200s。为了验证模型误差预测控制算法的有效性，这里针对两种不同情形进行了数值仿真。其中，情形一仅设计了 VSC 姿态控制器，对应的控制参数分别设置为 $\lambda=0.15$，$\varepsilon=0.1$，$G=0.015I_{3\times3}$；情形二则在情形一的基础上，进一步采用了模型误差预测控制算法补偿参数模型误差和外界扰动。为了避免 SGCMG 系统陷入奇异，当奇异量度 $D<0.1$ 时，采用 2.4.2.2 节中的奇异鲁棒操纵律，相关参数设置为 $\lambda_0=0.01$，$\mu=10$，$\varepsilon_0=0.01$，$\omega=1$，$\phi_i=\pi/2(i=1,2,3)$。仿真结果如图 6.20 和图 6.21 所示。

从图 6.20 和图 6.21 可以看出，由于存在较大外部干扰力矩，仅采用滑模变结构控制方法进行控制，存在较大姿态控制误差，几乎不能控；而采用

模型误差预测控制算法以后，能够很好地补偿参数模型误差和外界扰动，因此获得很好的控制精度，终端姿态控制精度不大于 0.01°，角速度控制精度不大于 0.001°/s。

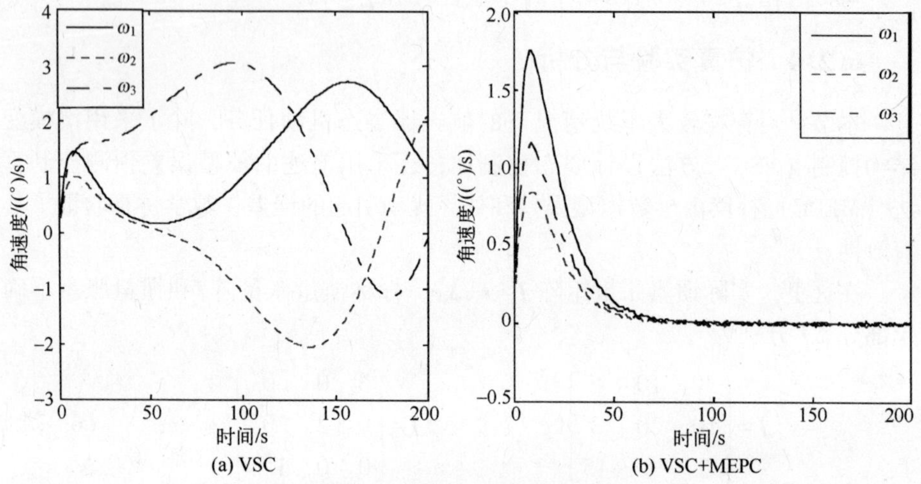

图 6.20　角速度变化曲线对比

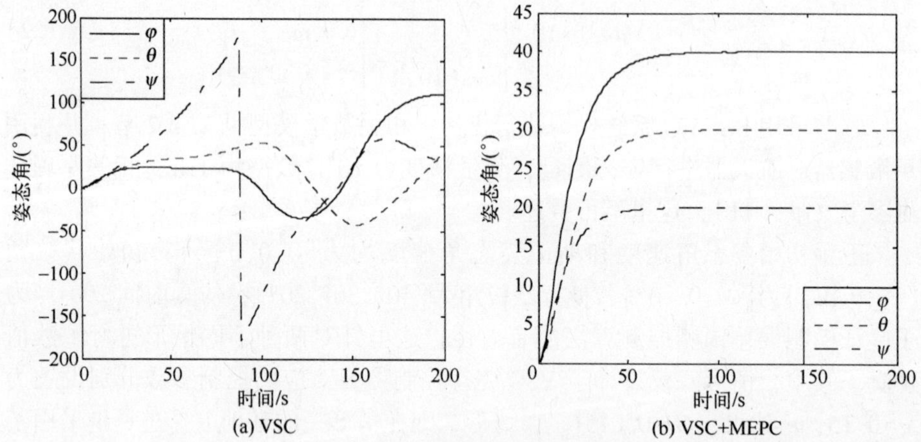

图 6.21　姿态角变化曲线对比

图 6.22 所示为外部干扰力矩估计曲线，可以看出，采用模型误差预测控制算法能够很好地对外部干扰力矩进行估计。图 6.23 所示为奇异量度变化曲线，图 6.24 和图 6.25 所示分别为框架角速度和框架角变化曲线。

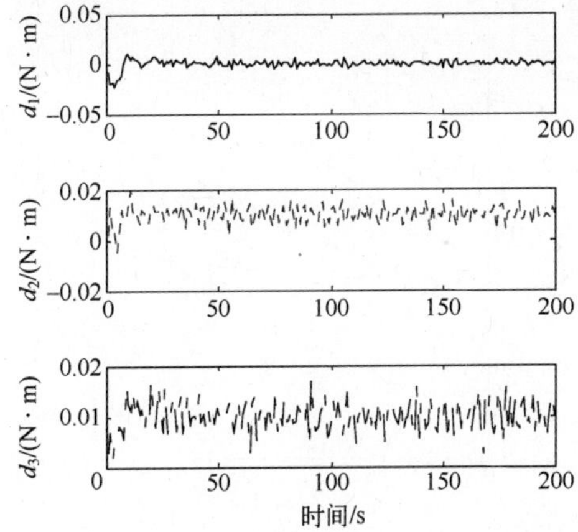

图 6.22　外部干扰力矩估计曲线（VSC+MEPC）

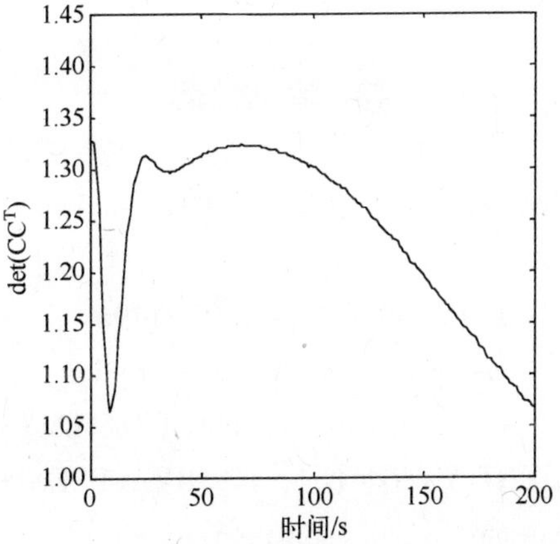

图 6.23　奇异量度变化曲线（VSC+MEPC）

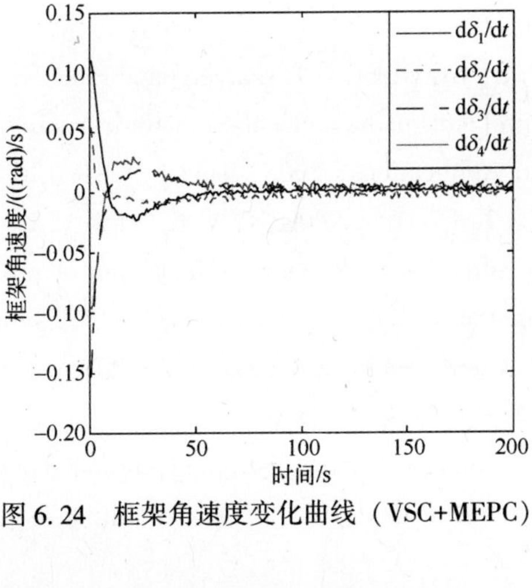

图 6.24　框架角速度变化曲线（VSC+MEPC）

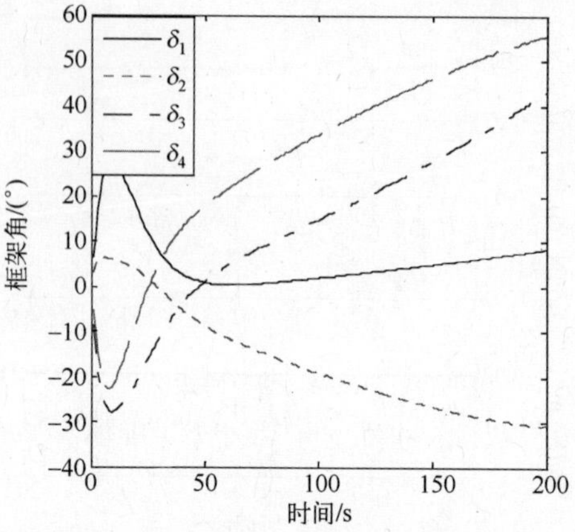

图 6.25　框架角变化曲线（VSC+MEPC）

参 考 文 献

[1] ELNAGAR G, KAZEMI M A, RAZZAGHI M. The pseudospectral legendre method for discretizing optimal control problems [J]. IEEE Transactions on Automatic Control, 1995, 40 (10): 1793-1796.

[2] BENSON D A. A Gauss pseudospectral transcription for optimal control [D]. Cambridge, M A: Massachusetts Institute of Technology, 2004.

[3] BENSON D A, HUNTINGTON G T, THORVALDSEN T P, et al. Direct trajectory optimization and costate estimation via an orthogonal collocation method [J]. Journal of Guidance, Control, and Dynamics, 2006, 29 (6): 1435-1440.

[4] KAMESWARAN S, BIEGLER L T. Convergence rates for direct transcription of optimal control problems using collocation at Radau points [J]. Computational Optimization and Applications, 2008, 41 (1): 81-126.

[5] GARG D, PATTERSON M A, DARBY C L, et al. Direct trajectory optimization and costate estimation of general optimal control problems using a Radau pseudospectral method [C] //AIAA Guidance, Navigation, and Control Conference, 10-13 August 2009, Chicago. USA: AIAA Inc, 2009: 1-29.

[6] FRANCOLIN C C, RAO A V. Direct trajectory optimization and costate estimation of state inequality path-constrained optimal control problems using a Radau pseudospectral method [C] // AIAA Guidance, Navigation, and Con-

trol Conference, 13-16 August 2012, Minneapolis, Minnesota. USA: AIAA Inc, 2012: 1-11.

[7] GARG D, PATTERSON M A, HAGER W W, et al. A unified framework for the numerical solution of optimal control problems using pseudospectral methods [J]. Automatica, 2010, 46 (11): 1843-1851.

[8] RAO A V, BENSON D A, DARBY C L, et al. User's manual for GPOPS version 5.0: a MATLAB software for solving multiple-phase optimal control problems using hp-adaptive pseudospectral methods [EB/OL]. [2011-08-01]. http://www.gpops.org/gpopsManual.pdf.

[9] CRASSIDIS J L, MARKLEY F L. Predictive filtering for nonlinear systems [J]. Journal of Guidance, Control, and Dynamics, 1997, 20 (3): 566-572.

[10] WIE B, BAILEY D, HEIBERG C. Rapid multi-target acquisition and pointing control of agile spacecraft [J]. Journal of Guidance, Control, and Dynamics, 2002, 25 (1): 96-104.

[11] 张力军. 基于多视场星敏感器的航天器姿态确定方法研究 [D]. 长沙: 国防科学技术大学, 2011.

附录 A　相对轨道动力学方程

A.1　任意相对轨道动力学方程

假设主星的地心矢径为 $\boldsymbol{r}_c$，从星的地心矢径为 $\boldsymbol{r}_d$，其相对主星的位置矢量为 $\boldsymbol{\rho}=\boldsymbol{r}_d-\boldsymbol{r}_c$。在地心惯性坐标系中，主星和从星的动力学方程为

$$\ddot{\boldsymbol{r}}_c=-\frac{\mu}{|\boldsymbol{r}_c|^3}\boldsymbol{r}_c+\boldsymbol{f}_{dc} \tag{A.1}$$

$$\ddot{\boldsymbol{r}}_d=-\frac{\mu}{|\boldsymbol{r}_d|^3}\boldsymbol{r}_d+\boldsymbol{a}_f+\boldsymbol{f}_{dd}=-\frac{\mu}{|\boldsymbol{r}_c+\boldsymbol{\rho}|^3}(\boldsymbol{r}_c+\boldsymbol{\rho})+\boldsymbol{a}_f+\boldsymbol{f}_{dd} \tag{A.2}$$

式中：μ 为地球引力常数；$\boldsymbol{f}_{dc}$ 和 $\boldsymbol{f}_{dd}$ 分别为主星和从星受到的所有摄动加速度；$\boldsymbol{a}_f$ 为从星上推力加速度矢量。将式（A.2）减去式（A.1）可得

$$\frac{\mathrm{d}^2\boldsymbol{\rho}}{\mathrm{d}t^2}=-\frac{\mu}{|\boldsymbol{r}_c+\boldsymbol{\rho}|^3}(\boldsymbol{r}_c+\boldsymbol{\rho})+\frac{\mu}{|\boldsymbol{r}_c|^3}\boldsymbol{r}_c+\boldsymbol{a}_f+\boldsymbol{f}_{dd}-\boldsymbol{f}_{dc} \tag{A.3}$$

另外，将相对加速度描述在主星 LVLH 系中可得

$$\frac{\mathrm{d}^2\boldsymbol{\rho}}{\mathrm{d}t^2}=\frac{\delta^2\boldsymbol{\rho}}{\delta t^2}+2\boldsymbol{\omega}\times\frac{\delta\boldsymbol{\rho}}{\delta t}+\dot{\boldsymbol{\omega}}\times\boldsymbol{\rho}+\boldsymbol{\omega}\times(\boldsymbol{\omega}\times\boldsymbol{\rho}) \tag{A.4}$$

式中：$\boldsymbol{\omega}$ 和 $\dot{\boldsymbol{\omega}}$ 分别为主星 LVLH 坐标系相对于地心惯性系的旋转角速度矢量与角加速度矢量。

联立式（A.3）和式（A.4）可得

$$\frac{\delta^2\boldsymbol{\rho}}{\delta t^2}=-\frac{\mu}{|\boldsymbol{r}_c+\boldsymbol{\rho}|^3}(\boldsymbol{r}_c+\boldsymbol{\rho})+\frac{\mu}{|\boldsymbol{r}_c|^3}\boldsymbol{r}_c-2\boldsymbol{\omega}\times\frac{\delta\boldsymbol{\rho}}{\delta t}-\dot{\boldsymbol{\omega}}\times\boldsymbol{\rho}-\boldsymbol{\omega}\times(\boldsymbol{\omega}\times\boldsymbol{\rho})+\boldsymbol{a}_f+\boldsymbol{f}_{dd}-\boldsymbol{f}_{dc} \tag{A.5}$$

该相对运动动力学方程适用于任意大的相对轨道以及主星轨道为椭圆轨道的情形。

A.2　10 维非线性相对运动动力学方程

对于主星和从星的近距离（几米至几十千米）相对运动情况，则相对轨

道坐标与主星的地心矢径相比是小量，可对式（A.5）右端的重力场进行一阶近似。具体如下：

$$|\boldsymbol{r}_c+\boldsymbol{\rho}|=[(\boldsymbol{r}_c+\boldsymbol{\rho})^{\mathrm{T}}(\boldsymbol{r}_c+\boldsymbol{\rho})]^{\frac{1}{2}}=(|\boldsymbol{r}_c|^2+2\boldsymbol{r}_c^{\mathrm{T}}\boldsymbol{\rho}+|\boldsymbol{\rho}|^2)^{\frac{1}{2}} \tag{A.6}$$

$$|\boldsymbol{r}_c+\boldsymbol{\rho}|^3=|\boldsymbol{r}_c|^3\left[1+2\frac{\boldsymbol{r}_c^{\mathrm{T}}\boldsymbol{\rho}}{|\boldsymbol{r}_c|^2}+\frac{|\boldsymbol{\rho}|^2}{|\boldsymbol{r}_c|^2}\right]^{\frac{3}{2}} \tag{A.7}$$

由于$|\boldsymbol{r}_c|\gg|\boldsymbol{\rho}|$，所以有

$$\begin{aligned}\frac{\boldsymbol{r}_c+\boldsymbol{\rho}}{|\boldsymbol{r}_c+\boldsymbol{\rho}|^3}&=\frac{\boldsymbol{r}_c+\boldsymbol{\rho}}{|\boldsymbol{r}_c|^3}\left(1+2\frac{\boldsymbol{r}_c^{\mathrm{T}}\boldsymbol{\rho}}{|\boldsymbol{r}_c|^2}+\frac{|\boldsymbol{\rho}|^2}{|\boldsymbol{r}_c|^2}\right)^{-\frac{3}{2}}\\&\approx\frac{\boldsymbol{r}_c+\boldsymbol{\rho}}{|\boldsymbol{r}_c|^3}\left[1-\frac{3}{2}\left(2\frac{\boldsymbol{r}_c^{\mathrm{T}}\boldsymbol{\rho}}{|\boldsymbol{r}_c|^2}+\frac{|\boldsymbol{\rho}|^2}{|\boldsymbol{r}_c|^2}\right)\right]\\&\approx\frac{1}{|\boldsymbol{r}_c|^3}\left(\boldsymbol{r}_c+\boldsymbol{\rho}-3\frac{\boldsymbol{r}_c^{\mathrm{T}}\boldsymbol{\rho}}{|\boldsymbol{r}_c|^2}\boldsymbol{r}_c\right)\end{aligned} \tag{A.8}$$

将式（A.8）代入式（A.5）可得线性化的相对轨道动力学方程为

$$\frac{\delta^2\boldsymbol{\rho}}{\delta t^2}=-\frac{\mu}{|\boldsymbol{r}_c|^3}\left(\boldsymbol{\rho}-3\frac{\boldsymbol{r}_c^{\mathrm{T}}\boldsymbol{\rho}}{|\boldsymbol{r}_c|^2}\boldsymbol{r}_c\right)-2\boldsymbol{\omega}\times\frac{\delta\boldsymbol{\rho}}{\delta t}-\dot{\boldsymbol{\omega}}\times\boldsymbol{\rho}-\boldsymbol{\omega}\times(\boldsymbol{\omega}\times\boldsymbol{\rho})+\boldsymbol{a}_f+\boldsymbol{f}_{dd}-\boldsymbol{f}_{dc} \tag{A.9}$$

根据坐标系定义，矢量$\boldsymbol{\omega}$，$\boldsymbol{r}_c$，$\boldsymbol{\rho}$在主星 LVLH 坐标系可分别表示成

$$\boldsymbol{\omega}=[0,0,\dot{\theta}]^{\mathrm{T}} \tag{A.10}$$

$$\boldsymbol{r}_c=[r_c,0,0]^{\mathrm{T}} \tag{A.11}$$

$$\boldsymbol{\rho}=[x,y,z]^{\mathrm{T}} \tag{A.12}$$

式中：r_c和$\dot{\theta}$分别为主星的地心距和真近点角速率。

令$\boldsymbol{\varpi}=\boldsymbol{f}_{dd}-\boldsymbol{f}_{dc}$，将式（A.10）、式（A.11）和式（A.12）代入式（A.9）中，并投影到主星 LVLH 系中可得相对运动动力学方程为

$$\begin{cases}\ddot{x}-x\left(\dot{\theta}^2+2\dfrac{\mu}{r_c^3}\right)-y\ddot{\theta}-2\dot{y}\dot{\theta}=a_x+\varpi_x\\ \ddot{y}+x\ddot{\theta}+2\dot{x}\dot{\theta}-y\left(\dot{\theta}^2-\dfrac{\mu}{r_c^3}\right)=a_y+\varpi_y\\ \ddot{z}+\dfrac{\mu}{r_c^3}z=a_z+\varpi_z\end{cases} \tag{A.13}$$

对于主星的轨道运动，有

$$\ddot{\theta}=-2\frac{\dot{r}_c}{r_c}\dot{\theta},\quad \ddot{r}_c=r_c\dot{\theta}^2-\frac{\mu}{r_c^2} \tag{A.14}$$

式中：$\ddot{\theta}$ 为主星的真近点角加速度大小。式（A.13）和式（A.14）组成了 10 维的非线性相对运动动力学方程。

若利用关系式 $\mu p = r_c^4 \dot{\theta}^2$，可将式（A.14）改写成用半通径 p 表示的相对运动动力学方程

$$\begin{cases} \ddot{x} - x\dot{\theta}^2\left(1+2\dfrac{r_c}{p}\right) - 2\dot{\theta}\left(\dot{y} - y\dfrac{\dot{r}_c}{r_c}\right) = a_x + \varpi_x \\ \ddot{y} + 2\dot{\theta}\left(\dot{x} - x\dfrac{\dot{r}_c}{r_c}\right) - y\dot{\theta}^2\left(1-\dfrac{r_c}{p}\right) = a_y + \varpi_y \\ \ddot{z} + z\dot{\theta}^2\dfrac{r_c}{p} = a_z + \varpi_z \end{cases} \tag{A.15}$$

$$\ddot{\theta} = -2\frac{\dot{r}_c}{r_c}\dot{\theta}, \quad \ddot{r}_c = r_c\dot{\theta}^2\left(1-\frac{r_c}{p}\right) \tag{A.16}$$

A.3 8 维非线性相对运动动力学方程

进一步，利用关系式

$$r_c^2\dot{\theta} = h \tag{A.17}$$

式中：h 为主星的轨道角动量，则可以定义常数 $k \equiv \mu/h^{\frac{3}{2}}$，有

$$\mu/r_c^3 = (\mu/h^{\frac{3}{2}})\dot{\theta}^{\frac{3}{2}} \equiv k\dot{\theta}^{\frac{3}{2}} \tag{A.18}$$

将式（A.18）代入式（A.13）可进一步得

$$\begin{cases} \ddot{x} - x(\dot{\theta}^2 + 2k\dot{\theta}^{3/2}) - y\ddot{\theta} - 2\dot{y}\dot{\theta} = a_x + \varpi_x \\ \ddot{y} + x\ddot{\theta} + 2\dot{x}\dot{\theta} - y(\dot{\theta}^2 - k\dot{\theta}^{3/2}) = a_y + \varpi_y \\ \ddot{z} + k\dot{\theta}^{3/2}z = a_z + \varpi_z \end{cases} \tag{A.19}$$

主星轨道角速度 $\dot{\theta}$ 和角加速度 $\ddot{\theta}$ 可分别表示成

$$\dot{\theta} = h/r_c^2 = \frac{1}{r_c}\sqrt{\frac{\mu}{p}}(1+e\cos\theta) = \frac{n(1+e\cos\theta)^2}{(1-e^2)^{3/2}} \tag{A.20}$$

$$\ddot{\theta} = \frac{-2ne\sin\theta(1+e\cos\theta)}{(1-e^2)^{3/2}}\dot{\theta} = \frac{-2n^2e\sin\theta(1+e\cos\theta)^3}{(1-e^2)^3} \tag{A.21}$$

式中：$n = \sqrt{\mu/a^3}$ 为主星的平均运动角速度；a 为主星轨道的半长轴。可以看出，方程（A.19）~方程（A.21）组成了一个 8 维的非线性微分方程组，相比式（A.13）和式（A.14）简化了 2 维。

A.4　C-W 方程

若主星轨道为圆轨道或近圆轨道，于是有 $e=0$，$\dot{r}_c=0$，$\ddot{\theta}=0$，则相对运动动力学方程就可简化为 C-W 方程：

$$\begin{cases} \ddot{x}-2n\dot{y}-3n^2x=a_x+\varpi_x \\ \ddot{y}+2n\dot{x}=a_y+\varpi_y \\ \ddot{z}+n^2z=a_z+\varpi_z \end{cases} \tag{A.22}$$

附录 B　偏导数矩阵$\partial \boldsymbol{f}(\boldsymbol{x}_p)/\partial \boldsymbol{x}_p$

状态矢量 $\boldsymbol{x}_p$ 包括相对位置、相对速度、主星的轨道半径、径向速率、真近点角以及真近点角速率，即

$$\begin{aligned}\boldsymbol{x}_p &= [x \quad y \quad z \quad \dot{x} \quad \dot{y} \quad \dot{z} \quad r_c \quad \dot{r}_c \quad \theta \quad \dot{\theta}]^{\mathrm{T}} \\ &\equiv [x_{p_1} \quad x_{p_2} \quad x_{p_3} \quad x_{p_4} \quad x_{p_5} \quad x_{p_6} \quad x_{p_7} \quad x_{p_8} \quad x_{p_9} \quad x_{p_{10}}]^{\mathrm{T}}\end{aligned} \tag{B.1}$$

式（4.32）中的偏导数矩阵$\partial \boldsymbol{f}(\boldsymbol{x}_p)/\partial \boldsymbol{x}_p$ 推导如下：

$$\left.\frac{\partial f(\boldsymbol{x}_p)}{\partial \boldsymbol{x}_p}\right|_{\hat{x}_p} = \begin{bmatrix} 0 & 0 & 0 & 1 & 0 & 0 & 0 & 0 & 0 & 0 \\ 0 & 0 & 0 & 0 & 1 & 0 & 0 & 0 & 0 & 0 \\ 0 & 0 & 0 & 0 & 0 & 1 & 0 & 0 & 0 & 0 \\ J_1 & -2\dfrac{x_{p_8}}{x_{p_7}}x_{p_{10}} & 0 & 0 & 2x_{p_{10}} & 0 & J_2 & -2\dfrac{x_{p_2}}{x_{p_7}}x_{p_{10}} & 0 & J_3 \\ 2\dfrac{x_{p_8}}{x_{p_7}}x_{p_{10}} & J_4 & 0 & -2x_{p_{10}} & 0 & 0 & J_5 & 2\dfrac{x_{p_1}}{x_{p_7}}x_{p_{10}} & 0 & J_6 \\ 0 & 0 & -\dfrac{x_{p_7}}{p}x_{p_{10}}^2 & 0 & 0 & 0 & -\dfrac{1}{p}x_{p_{10}}^2 x_{p_3} & 0 & 0 & J_7 \\ 0 & 0 & 0 & 0 & 0 & 0 & 0 & 1 & 0 & 0 \\ 0 & 0 & 0 & 0 & 0 & 0 & J_8 & 0 & 0 & J_9 \\ 0 & 0 & 0 & 0 & 0 & 0 & 0 & 0 & 0 & 1 \\ 0 & 0 & 0 & 0 & 0 & 0 & 2\dfrac{x_{p_8}}{x_{p_7}^2}x_{p_{10}} & -\dfrac{2}{x_{p_7}}x_{p_{10}} & 0 & -2\dfrac{x_{p_8}}{x_{p_7}} \end{bmatrix} \tag{B.2}$$

其中，

$$
\begin{cases}
J_1 = x_{p_{10}}^2 (1+2x_{p_7}/p) \\
J_2 = 2x_{p_1} x_{p_{10}}^2 /p + 2x_{p_{10}} x_{p_2} x_{p_8} / x_7^2 \\
J_3 = 2x_{p_1} x_{p_{10}} (1+2x_{p_7}/p) + 2x_{p_5} - 2x_{p_2} x_{p_8} / x_{p_7} \\
J_4 = x_{p_{10}}^2 (1-x_{p_7}/p) \\
J_5 = -2x_{p_{10}} x_{p_1} x_{p_8} / x_{p_7}^2 - x_{p_2} x_{p_{10}}^2 /p \\
J_6 = -2x_{p_4} + 2x_{p_1} x_{p_8} / x_{p_7} + 2x_{p_2} x_{p_{10}} (1-x_{p_7}/p) \\
J_7 = -2x_{p_{10}} x_{p_7} x_{p_3} /p \\
J_8 = x_{10}^2 (1-x_{p_7}/p) - x_{10}^2 x_{p_7} /p \\
J_9 = 2x_{p_7} x_{p_{10}} (1-x_{p_7}/p)
\end{cases}
\quad (\mathrm{B.3})
$$

式中：p 为主星轨道的半通径。

附录 C　第 5 章相关公式推导

对于两个三维矢量 $\boldsymbol{v}_1$ 和 $\boldsymbol{v}_2$ 而言，由四元数乘法规则可知它们间的矢量积可用对应的广义四元数表示为

$$\begin{aligned}\boldsymbol{v}_1\times\boldsymbol{v}_2&=\frac{1}{2}(\bar{\boldsymbol{v}}_2\otimes\bar{\boldsymbol{v}}_1-\bar{\boldsymbol{v}}_1\otimes\bar{\boldsymbol{v}}_2)\\&=\frac{1}{2}(\bar{\boldsymbol{v}}_1\otimes\bar{\boldsymbol{v}}_2^*-\bar{\boldsymbol{v}}_2\otimes\bar{\boldsymbol{v}}_1^*)\end{aligned}\tag{C.1}$$

式（C.1）将在后续公式推导中反复用到。

对式（5.22）两边同时求导可得

$$\begin{aligned}\dot{\boldsymbol{p}}_{d/t}^{d}&=\dot{\boldsymbol{r}}_d^d-\dot{\boldsymbol{q}}_e\otimes\boldsymbol{r}_t^t\otimes\boldsymbol{q}_e^*-\boldsymbol{q}_e\otimes\dot{\boldsymbol{r}}_t^t\otimes\boldsymbol{q}_e^*-\boldsymbol{q}_e\otimes\boldsymbol{r}_t^t\otimes\dot{\boldsymbol{q}}_e^*\\&=\dot{\boldsymbol{r}}_d^d-\boldsymbol{q}_e\otimes\dot{\boldsymbol{r}}_t^t\otimes\boldsymbol{q}_e^*-\frac{1}{2}\boldsymbol{\omega}_{d/t}^d\otimes\boldsymbol{q}_e\otimes\boldsymbol{r}_t^t\otimes\boldsymbol{q}_e^*+\frac{1}{2}\boldsymbol{q}_e\otimes\boldsymbol{r}_t^t\otimes\boldsymbol{q}_e^*\otimes\boldsymbol{\omega}_{d/t}^d\\&=\dot{\boldsymbol{r}}_d^d-\boldsymbol{q}_e\otimes\dot{\boldsymbol{r}}_t^t\otimes\boldsymbol{q}_e^*+\boldsymbol{\omega}_{d/t}^d\times(\boldsymbol{q}_e\otimes\boldsymbol{r}_t^t\otimes\boldsymbol{q}_e^*)\\&=\dot{\boldsymbol{r}}_d^d-\boldsymbol{q}_e\otimes\dot{\boldsymbol{r}}_t^t\otimes\boldsymbol{q}_e^*+\boldsymbol{\omega}_{d/t}^d\times\boldsymbol{r}_t^d\end{aligned}\tag{C.2}$$

C.1　式（5.26）的推导

由式（5.26）可知，d 系相对于 t 系的误差旋量 $\widehat{\boldsymbol{\omega}}_{d/t}^d$ 为

$$\begin{aligned}\widehat{\boldsymbol{\omega}}_{d/t}^d&=\widehat{\boldsymbol{\omega}}_{d/I}^d-\widehat{\boldsymbol{q}}_e\otimes\widehat{\boldsymbol{\omega}}_{t/I}^t\otimes\widehat{\boldsymbol{q}}_e^*\\&=\boldsymbol{\omega}_{d/I}^d+\varepsilon(\dot{\boldsymbol{r}}_d^d+\boldsymbol{\omega}_{d/I}^d\times\boldsymbol{r}_d^d)-\left(\boldsymbol{q}_e+\varepsilon\frac{1}{2}\boldsymbol{p}_{d/t}^d\otimes\boldsymbol{q}_e\right)\otimes(\boldsymbol{\omega}_{t/I}^t+\varepsilon(\dot{\boldsymbol{r}}_t^t+\boldsymbol{\omega}_{t/I}^t\times\boldsymbol{r}_t^t))\otimes\left(\boldsymbol{q}_e^*-\varepsilon\frac{1}{2}\boldsymbol{q}_e^*\otimes\boldsymbol{p}_{d/t}^d\right)\\&=(\boldsymbol{\omega}_{d/I}^d-\boldsymbol{q}_e\otimes\boldsymbol{\omega}_{t/I}^t\otimes\boldsymbol{q}_e^*)\\&\quad+\varepsilon(\dot{\boldsymbol{r}}_d^d+\boldsymbol{\omega}_{d/I}^d\times\boldsymbol{r}_d^d-\boldsymbol{q}_e\otimes(\dot{\boldsymbol{r}}_t^t+\boldsymbol{\omega}_{t/I}^t\times\boldsymbol{r}_t^t)\otimes\boldsymbol{q}_e^*-\frac{1}{2}\boldsymbol{p}_{d/t}^d\otimes\boldsymbol{q}_e\otimes\boldsymbol{\omega}_{t/I}^t\otimes\boldsymbol{q}_e^*+\frac{1}{2}\boldsymbol{q}_e\otimes\boldsymbol{\omega}_{t/I}^t\otimes\boldsymbol{q}_e^*\otimes\boldsymbol{p}_{d/t}^d)\\&=\boldsymbol{\omega}_{d/t}^d+\varepsilon(\dot{\boldsymbol{r}}_d^d-\boldsymbol{q}_e\otimes\dot{\boldsymbol{r}}_t^t\otimes\boldsymbol{q}_e^*+\boldsymbol{\omega}_{d/I}^d\times\boldsymbol{r}_d^d-\boldsymbol{\omega}_{t/I}^d\times\boldsymbol{r}_t^d-\boldsymbol{\omega}_{t/I}^d\times\boldsymbol{p}_{d/t}^d)\\&=\boldsymbol{\omega}_{d/t}^d+\varepsilon(\dot{\boldsymbol{r}}_d^d-\boldsymbol{q}_e\otimes\dot{\boldsymbol{r}}_t^t\otimes\boldsymbol{q}_e^*+\boldsymbol{\omega}_{d/I}^d\times\boldsymbol{r}_d^d-\boldsymbol{\omega}_{t/I}^d\times\boldsymbol{r}_d^d)\\&=\boldsymbol{\omega}_{d/t}^d+\varepsilon(\dot{\boldsymbol{r}}_d^d-\boldsymbol{q}_e\otimes\dot{\boldsymbol{r}}_t^t\otimes\boldsymbol{q}_e^*+\boldsymbol{\omega}_{d/t}^d\times(\boldsymbol{p}_{d/t}^d+\boldsymbol{r}_t^d))\\&=\boldsymbol{\omega}_{d/t}^d+\varepsilon(\dot{\boldsymbol{r}}_d^d-\boldsymbol{q}_e\otimes\dot{\boldsymbol{r}}_t^t\otimes\boldsymbol{q}_e^*+\boldsymbol{\omega}_{d/t}^d\times\boldsymbol{r}_t^d+\boldsymbol{\omega}_{d/t}^d\times\boldsymbol{p}_{d/t}^d)\\&=\boldsymbol{\omega}_{d/t}^d+\varepsilon(\dot{\boldsymbol{p}}_{d/t}^d+\boldsymbol{\omega}_{d/t}^d\times\boldsymbol{p}_{d/t}^d)\end{aligned}\tag{C.3}$$

C.2　式（5.41）和式（5.42）的推导

将式（5.40）进一步展开分为实部和对偶部，可分别得到相对转动误差动力学方程和相对平动误差动力学方程，详细推导如下所述。

对式（C.3）两边同时求导，可得式（5.40）的左边为

$$\dot{\hat{\boldsymbol{\omega}}}_{d/t}^{d}=\dot{\boldsymbol{\omega}}_{d/t}^{d}+\varepsilon(\ddot{\boldsymbol{p}}_{d/t}^{d}+\dot{\boldsymbol{\omega}}_{d/t}^{d}\times\boldsymbol{p}_{d/t}^{d}+\boldsymbol{\omega}_{d/t}^{d}\times\dot{\boldsymbol{p}}_{d/t}^{d}) \tag{C.4}$$

对于式（5.40）的右边，将对每一项先逐项展开分为实部和对偶部，然后合并同类项并进行化简。

1. 逐项展开

式（5.40）右边第一项展开为

$$\begin{aligned}
-\widehat{\boldsymbol{M}}_{d}^{-1}(\hat{\boldsymbol{\omega}}_{d/I}^{d}\times\widehat{\boldsymbol{M}}_{d}\,\hat{\boldsymbol{\omega}}_{d/I}^{d})&=-\left(\boldsymbol{J}_{d}^{-1}\frac{d}{d\varepsilon}+\varepsilon\frac{1}{m_{d}}\boldsymbol{I}_{3\times3}\right)[(\boldsymbol{\omega}_{d/I}^{d}+\varepsilon\boldsymbol{v}_{d}^{d})\times(m_{d}\boldsymbol{v}_{d}^{d}+\varepsilon\boldsymbol{J}_{d}\boldsymbol{\omega}_{d/I}^{d})]\\
&=-\left(\boldsymbol{J}_{d}^{-1}\frac{d}{d\varepsilon}+\varepsilon\frac{1}{m_{d}}\boldsymbol{I}_{3\times3}\right)[m_{d}\boldsymbol{\omega}_{d/I}^{d}\times\boldsymbol{v}_{d}^{d}+\varepsilon(\boldsymbol{\omega}_{d/I}^{d}\times\boldsymbol{J}_{d}\boldsymbol{\omega}_{d/I}^{d})]\\
&=-\boldsymbol{J}_{d}^{-1}(\boldsymbol{\omega}_{d/I}^{d}\times\boldsymbol{J}_{d}\boldsymbol{\omega}_{d/I}^{d})-\varepsilon(\boldsymbol{\omega}_{d/I}^{d}\times\boldsymbol{v}_{d}^{d})
\end{aligned} \tag{C.5}$$

对于$-(\boldsymbol{\omega}_{d/I}^{d}\times\boldsymbol{v}_{d}^{d})$，依据定义可进一步展开为

$$-(\boldsymbol{\omega}_{d/I}^{d}\times\boldsymbol{v}_{d}^{d})=-\boldsymbol{\omega}_{d/I}^{d}\times(\dot{\boldsymbol{r}}_{d}^{d}+\boldsymbol{\omega}_{d/I}^{d}\times\boldsymbol{r}_{d}^{d})=-\boldsymbol{\omega}_{d/I}^{d}\times[\dot{\boldsymbol{p}}_{d/t}^{d}+\boldsymbol{A}(\boldsymbol{q}_{e})\dot{\boldsymbol{r}}_{t}^{t}-\boldsymbol{\omega}_{d/t}^{d}\times\boldsymbol{r}_{t}^{d}+\boldsymbol{\omega}_{d/I}^{d}\times(\boldsymbol{p}_{d/t}^{d}+\boldsymbol{r}_{t}^{d})] \tag{C.6}$$

第二项展开为

$$\begin{aligned}
\widehat{\boldsymbol{M}}_{d}^{-1}(\hat{\boldsymbol{u}}_{c}^{d}+\hat{\boldsymbol{u}}_{g}^{d}+\hat{\boldsymbol{d}})&=\left(\boldsymbol{J}_{d}^{-1}\frac{d}{d\varepsilon}+\varepsilon\frac{1}{m_{d}}\boldsymbol{I}_{3\times3}\right)[(\boldsymbol{f}_{c}^{d}+\boldsymbol{f}_{g}^{d}+\boldsymbol{f}_{d})+\varepsilon(\boldsymbol{\tau}_{c}^{d}+\boldsymbol{\tau}_{g}^{d}+\boldsymbol{\tau}_{d})]\\
&=\boldsymbol{J}_{d}^{-1}(\boldsymbol{\tau}_{c}^{d}+\boldsymbol{\tau}_{g}^{d}+\boldsymbol{\tau}_{d})+\varepsilon\frac{1}{m_{d}}(\boldsymbol{f}_{c}^{d}+\boldsymbol{f}_{g}^{d}+\boldsymbol{f}_{d})
\end{aligned} \tag{C.7}$$

第三项展开为

$$\begin{aligned}
&-\hat{\boldsymbol{q}}_{e}\otimes\dot{\hat{\boldsymbol{\omega}}}_{t/I}^{t}\otimes\hat{\boldsymbol{q}}_{e}^{*}\\
&=-\left(\boldsymbol{q}_{e}+\varepsilon\frac{1}{2}\boldsymbol{p}_{d/t}^{d}\otimes\boldsymbol{q}_{e}\right)\otimes(\dot{\boldsymbol{\omega}}_{t/I}^{t}+\varepsilon(\ddot{\boldsymbol{r}}_{t}^{t}+\dot{\boldsymbol{\omega}}_{t/I}^{t}\times\boldsymbol{r}_{t}^{t}+\boldsymbol{\omega}_{t/I}^{t}\times\dot{\boldsymbol{r}}_{t}^{t}))\otimes\left(\boldsymbol{q}_{e}^{*}-\varepsilon\frac{1}{2}\boldsymbol{q}_{e}^{*}\otimes\boldsymbol{p}_{d/t}^{d}\right)\\
&=-\boldsymbol{A}(\boldsymbol{q}_{e})\dot{\boldsymbol{\omega}}_{t/I}^{t}-\varepsilon[\boldsymbol{A}(\boldsymbol{q}_{e})\dot{\boldsymbol{\omega}}_{t/I}^{t}\times\boldsymbol{p}_{d/t}^{d}+\boldsymbol{A}(\boldsymbol{q}_{e})(\ddot{\boldsymbol{r}}_{t}^{t}+\dot{\boldsymbol{\omega}}_{t/I}^{t}\times\boldsymbol{r}_{t}^{t}+\boldsymbol{\omega}_{t/I}^{t}\times\dot{\boldsymbol{r}}_{t}^{t})]
\end{aligned} \tag{C.8}$$

对于第四项，首先对$\hat{\boldsymbol{q}}_{e}\otimes\hat{\boldsymbol{\omega}}_{t/I}^{t}\otimes\hat{\boldsymbol{q}}_{e}^{*}$进行展开，可得

$$\hat{\boldsymbol{q}}_e \otimes \hat{\boldsymbol{\omega}}_{t/I}^t \otimes \hat{\boldsymbol{q}}_e^* = \left(\boldsymbol{q}_e + \varepsilon \times \frac{1}{2}\boldsymbol{p}_{d/t}^d \otimes \boldsymbol{q}_e\right) \otimes (\boldsymbol{\omega}_{t/I}^t + \varepsilon(\dot{\boldsymbol{r}}_t^t + \boldsymbol{\omega}_{t/I}^t \times \boldsymbol{r}_t^t)) \otimes \left(\boldsymbol{q}_e^* - \varepsilon \times \frac{1}{2}\boldsymbol{q}_e^* \otimes \boldsymbol{p}_{d/t}^d\right)$$
$$= \boldsymbol{A}(\boldsymbol{q}_e)\boldsymbol{\omega}_{t/I}^t + \varepsilon[\boldsymbol{A}(\boldsymbol{q}_e)\boldsymbol{\omega}_{t/I}^t \times \boldsymbol{p}_{d/t}^d + \boldsymbol{A}(\boldsymbol{q}_e)(\dot{\boldsymbol{r}}_t^t + \boldsymbol{\omega}_{t/I}^t \times \boldsymbol{r}_t^t)] \tag{C.9}$$

因此，第四项展开为

$$\hat{\boldsymbol{\omega}}_{d/t}^d \times (\hat{\boldsymbol{q}}_e \otimes \hat{\boldsymbol{\omega}}_{t/I}^t \otimes \hat{\boldsymbol{q}}_e^*)$$
$$= [\boldsymbol{\omega}_{d/t}^d + \varepsilon(\dot{\boldsymbol{p}}_{d/t}^d + \boldsymbol{\omega}_{d/t}^d \times \boldsymbol{p}_{d/t}^d)] \times \{\boldsymbol{A}(\boldsymbol{q}_e)\boldsymbol{\omega}_{t/I}^t + \varepsilon[\boldsymbol{A}(\boldsymbol{q}_e)\boldsymbol{\omega}_{t/I}^t \times \boldsymbol{p}_{d/t}^d + \boldsymbol{A}(\boldsymbol{q}_e)(\dot{\boldsymbol{r}}_t^t + \boldsymbol{\omega}_{t/I}^t \times \boldsymbol{r}_t^t)]\}$$
$$= \boldsymbol{\omega}_{d/t}^d \times \boldsymbol{A}(\boldsymbol{q}_e)\boldsymbol{\omega}_{t/I}^t + \varepsilon[\boldsymbol{\omega}_{d/t}^d \times (\boldsymbol{A}(\boldsymbol{q}_e)\boldsymbol{\omega}_{t/I}^t \times \boldsymbol{p}_{d/t}^d) + \boldsymbol{\omega}_{d/t}^d \times \boldsymbol{A}(\boldsymbol{q}_e)(\dot{\boldsymbol{r}}_t^t + \boldsymbol{\omega}_{t/I}^t \times \boldsymbol{r}_t^t) +$$
$$(\dot{\boldsymbol{p}}_{d/t}^d + \boldsymbol{\omega}_{d/t}^d \times \boldsymbol{p}_{d/t}^d) \times \boldsymbol{A}(\boldsymbol{q}_e)\boldsymbol{\omega}_{t/I}^t] \tag{C.10}$$

利用雅克比恒等式

$$(\boldsymbol{a} \times \boldsymbol{b}) \times \boldsymbol{c} + (\boldsymbol{b} \times \boldsymbol{c}) \times \boldsymbol{a} + (\boldsymbol{c} \times \boldsymbol{a}) \times \boldsymbol{b} = \boldsymbol{0}_{3\times1} \tag{C.11}$$

可对式（C.10）的对偶部进行部分化简，即有

$$\boldsymbol{\omega}_{d/t}^d \times (\boldsymbol{A}(\boldsymbol{q}_e)\boldsymbol{\omega}_{t/I}^t \times \boldsymbol{p}_{d/t}^d) + (\boldsymbol{\omega}_{d/t}^d \times \boldsymbol{p}_{d/t}^d) \times \boldsymbol{A}(\boldsymbol{q}_e)\boldsymbol{\omega}_{t/I}^t$$
$$= -(\boldsymbol{A}(\boldsymbol{q}_e)\boldsymbol{\omega}_{t/I}^t \times \boldsymbol{p}_{d/t}^d) \times \boldsymbol{\omega}_{d/t}^d - (\boldsymbol{p}_{d/t}^d \times \boldsymbol{\omega}_{d/t}^d) \times \boldsymbol{A}(\boldsymbol{q}_e)\boldsymbol{\omega}_{t/I}^t$$
$$= (\boldsymbol{\omega}_{d/t}^d \times \boldsymbol{A}(\boldsymbol{q}_e)\boldsymbol{\omega}_{t/I}^t) \times \boldsymbol{p}_{d/t}^d$$
$$= -(\boldsymbol{A}(\boldsymbol{q}_e)\boldsymbol{\omega}_{t/I}^t \times \boldsymbol{\omega}_{d/t}^d) \times \boldsymbol{p}_{d/t}^d \tag{C.12}$$

因此，式（C.10）可改写为

$$\hat{\boldsymbol{\omega}}_{d/t}^d \times (\hat{\boldsymbol{q}}_e \otimes \hat{\boldsymbol{\omega}}_{t/I}^t \otimes \hat{\boldsymbol{q}}_e^*)$$
$$= \boldsymbol{\omega}_{d/t}^d \times \boldsymbol{A}(\boldsymbol{q}_e)\boldsymbol{\omega}_{t/I}^t + \varepsilon[\boldsymbol{\omega}_{d/t}^d \times \boldsymbol{A}(\boldsymbol{q}_e)(\dot{\boldsymbol{r}}_t^t + \boldsymbol{\omega}_{t/I}^t \times \boldsymbol{r}_t^t) - (\boldsymbol{A}(\boldsymbol{q}_e)\boldsymbol{\omega}_{t/I}^t \times \boldsymbol{\omega}_{d/t}^d) \times \boldsymbol{p}_{d/t}^d - \boldsymbol{A}(\boldsymbol{q}_e)\boldsymbol{\omega}_{t/I}^t \times \dot{\boldsymbol{p}}_{d/t}^d] \tag{C.13}$$

2. 合并同类项并化简

利用式（C.4）~式（C.13），通过合并同类项和化简，可将式（5.40）展开分为实部和对偶部。

对应的实部部分为

$$\dot{\boldsymbol{\omega}}_{d/t}^d = -\boldsymbol{J}_d^{-1}(\boldsymbol{\omega}_{d/I}^d \times \boldsymbol{J}_d \boldsymbol{\omega}_{d/I}^d) - (\boldsymbol{A}(\boldsymbol{q}_e)\dot{\boldsymbol{\omega}}_{t/I}^t - \boldsymbol{\omega}_{d/t}^d \times \boldsymbol{A}(\boldsymbol{q}_e)\boldsymbol{\omega}_{t/I}^t) + \boldsymbol{J}_d^{-1}(\boldsymbol{\tau}_c^d + \boldsymbol{\tau}_g^d + \boldsymbol{\tau}_d) \tag{C.14}$$

式（C.14）即相对转动误差动力学方程，与传统的姿态跟踪误差动力学方程一致。

对偶部部分为

$$
\begin{aligned}
\ddot{\boldsymbol{p}}_{d/t}^{d} &= -\dot{\boldsymbol{\omega}}_{d/t}^{d}\times\boldsymbol{p}_{d/t}^{d}-\boldsymbol{\omega}_{d/t}^{d}\times\dot{\boldsymbol{p}}_{d/t}^{d}- \\
&\quad \boldsymbol{\omega}_{d/I}^{d}\times[\dot{\boldsymbol{p}}_{d/t}^{d}+\boldsymbol{A}(\boldsymbol{q}_e)\dot{\boldsymbol{r}}_t^{t}-\boldsymbol{\omega}_{d/t}^{d}\times\boldsymbol{r}_t^{d}+\boldsymbol{\omega}_{d/I}^{d}\times(\boldsymbol{p}_{d/t}^{d}+\boldsymbol{r}_t^{d})]+ \\
&\quad \frac{1}{m_d}(\boldsymbol{f}_c^{d}+\boldsymbol{f}_g^{d}+\boldsymbol{f}_d)- \\
&\quad [\boldsymbol{A}(\boldsymbol{q}_e)\dot{\boldsymbol{\omega}}_{t/I}^{t}\times\boldsymbol{p}_{d/t}^{d}+\boldsymbol{A}(\boldsymbol{q}_e)(\ddot{\boldsymbol{r}}_t^{t}+\dot{\boldsymbol{\omega}}_{t/I}^{t}\times\boldsymbol{r}_t^{t}+\boldsymbol{\omega}_{t/I}^{t}\times\dot{\boldsymbol{r}}_t^{t})]+ \\
&\quad [\boldsymbol{\omega}_{d/t}^{d}\times\boldsymbol{A}(\boldsymbol{q}_e)(\dot{\boldsymbol{r}}_t^{t}+\boldsymbol{\omega}_{t/I}^{t}\times\boldsymbol{r}_t^{t})-(\boldsymbol{A}(\boldsymbol{q}_e)\boldsymbol{\omega}_{t/I}^{t}\times\boldsymbol{\omega}_{d/t}^{d})\times\boldsymbol{p}_{d/t}^{d}-\boldsymbol{A}(\boldsymbol{q}_e)\boldsymbol{\omega}_{t/I}^{t}\times\dot{\boldsymbol{p}}_{d/t}^{d}] \\
&= -[\dot{\boldsymbol{\omega}}_{d/t}^{d}+\boldsymbol{A}(\boldsymbol{q}_e)\boldsymbol{\omega}_{t/I}^{t}\times\boldsymbol{\omega}_{d/t}^{d}+\boldsymbol{A}(\boldsymbol{q}_e)\dot{\boldsymbol{\omega}}_{t/I}^{t}]\times\boldsymbol{p}_{d/t}^{d}-\boldsymbol{\omega}_{d/I}^{d}\times(\boldsymbol{\omega}_{d/I}^{d}\times\boldsymbol{p}_{d/t}^{d})- \\
&\quad (\boldsymbol{\omega}_{d/I}^{d}+\boldsymbol{\omega}_{d/t}^{d}+\boldsymbol{A}(\boldsymbol{q}_e)\boldsymbol{\omega}_{t/I}^{t})\times\dot{\boldsymbol{p}}_{d/t}^{d}-\frac{\mu}{r_d^3}\boldsymbol{r}_d^{d}+\frac{\boldsymbol{f}_c^{d}}{m_d}+\frac{\boldsymbol{f}_d}{m_d}- \\
&\quad \{\boldsymbol{A}(\boldsymbol{q}_e)(\ddot{\boldsymbol{r}}_t^{t}+\dot{\boldsymbol{\omega}}_{t/I}^{t}\times\boldsymbol{r}_t^{t}+\boldsymbol{\omega}_{t/I}^{t}\times\dot{\boldsymbol{r}}_t^{t})+(\boldsymbol{\omega}_{d/I}^{d}-\boldsymbol{\omega}_{d/t}^{d})\times\boldsymbol{A}(\boldsymbol{q}_e)\dot{\boldsymbol{r}}_t^{t}+\boldsymbol{\omega}_{d/I}^{d}\times[(\boldsymbol{\omega}_{d/I}^{d}-\boldsymbol{\omega}_{d/t}^{d})\times\boldsymbol{r}_t^{d}] \\
&\quad -\boldsymbol{\omega}_{d/t}^{d}\times\boldsymbol{A}(\boldsymbol{q}_e)(\boldsymbol{\omega}_{t/I}^{t}\times\boldsymbol{r}_t^{t})\}
\end{aligned}
\tag{C.15}
$$

对于式（C.15）中右侧大括号项，将其进一步化简可得

$$
\begin{aligned}
&\boldsymbol{A}(\boldsymbol{q}_e)(\ddot{\boldsymbol{r}}_t^{t}+\dot{\boldsymbol{\omega}}_{t/I}^{t}\times\boldsymbol{r}_t^{t}+\boldsymbol{\omega}_{t/I}^{t}\times\dot{\boldsymbol{r}}_t^{t})+(\boldsymbol{\omega}_{d/I}^{d}-\boldsymbol{\omega}_{d/t}^{d})\times\boldsymbol{A}(\boldsymbol{q}_e)\dot{\boldsymbol{r}}_t^{t}+\boldsymbol{\omega}_{d/I}^{d}\times \\
&\quad [(\boldsymbol{\omega}_{d/I}^{d}-\boldsymbol{\omega}_{d/t}^{d})\times\boldsymbol{r}_t^{d}]-\boldsymbol{\omega}_{d/t}^{d}\times\boldsymbol{A}(\boldsymbol{q}_e)(\boldsymbol{\omega}_{t/I}^{t}\times\boldsymbol{r}_t^{t}) \\
&=\boldsymbol{A}(\boldsymbol{q}_e)(\ddot{\boldsymbol{r}}_t^{t}+\dot{\boldsymbol{\omega}}_{t/I}^{t}\times\boldsymbol{r}_t^{t}+\boldsymbol{\omega}_{t/I}^{t}\times\dot{\boldsymbol{r}}_t^{t})+\boldsymbol{A}(\boldsymbol{q}_e)\boldsymbol{\omega}_{t/I}^{t}\times\boldsymbol{A}(\boldsymbol{q}_e)\dot{\boldsymbol{r}}_t^{t}+(\boldsymbol{\omega}_{d/I}^{d}-\boldsymbol{\omega}_{d/t}^{d})\times\boldsymbol{A}(\boldsymbol{q}_e)(\boldsymbol{\omega}_{t/I}^{t}\times\boldsymbol{r}_t^{t}) \\
&=\boldsymbol{A}(\boldsymbol{q}_e)(\ddot{\boldsymbol{r}}_t^{t}+\dot{\boldsymbol{\omega}}_{t/I}^{t}\times\boldsymbol{r}_t^{t}+\boldsymbol{\omega}_{t/I}^{t}\times\dot{\boldsymbol{r}}_t^{t})+\boldsymbol{A}(\boldsymbol{q}_e)(\boldsymbol{\omega}_{t/I}^{t}\times\dot{\boldsymbol{r}}_t^{t})+\boldsymbol{A}(\boldsymbol{q}_e)[\boldsymbol{\omega}_{t/I}^{t}\times(\boldsymbol{\omega}_{t/I}^{t}\times\boldsymbol{r}_t^{t})] \\
&=\boldsymbol{A}(\boldsymbol{q}_e)(\ddot{\boldsymbol{r}}_t^{t}+2\boldsymbol{\omega}_{t/I}^{t}\times\dot{\boldsymbol{r}}_t^{t}+\dot{\boldsymbol{\omega}}_{t/I}^{t}\times\boldsymbol{r}_t^{t}+\boldsymbol{\omega}_{t/I}^{t}\times(\boldsymbol{\omega}_{t/I}^{t}\times\boldsymbol{r}_t^{t}))
\end{aligned}
\tag{C.16}
$$

对于期望坐标系的地心惯性矢径 $\boldsymbol{r}_t$，其在地心惯性坐标系中的动力学方程可描述为

$$
\ddot{\boldsymbol{r}}_t=-\frac{\mu}{r_t^3}\boldsymbol{r}_t \tag{C.17}
$$

式中：μ 为地球引力常数；r_t 为地心距。

参照附录 A 中的推导，可将式（C.17）描述在期望坐标系 t 系中，可得

$$
\ddot{\boldsymbol{r}}_t^{t}+2\boldsymbol{\omega}_{t/I}^{t}\times\dot{\boldsymbol{r}}_t^{t}+\dot{\boldsymbol{\omega}}_{t/I}^{t}\times\boldsymbol{r}_t^{t}+\boldsymbol{\omega}_{t/I}^{t}\times(\boldsymbol{\omega}_{t/I}^{t}\times\boldsymbol{r}_t^{t})=-\frac{\mu}{r_t^3}\boldsymbol{r}_t^{t} \tag{C.18}
$$

因此，结合式（C.16）和式（C.18），可进一步将式（C.15）改写为

$$
\begin{aligned}
\ddot{\boldsymbol{p}}_{d/t}^{d} &= -[\dot{\boldsymbol{\omega}}_{d/t}^{d}+\boldsymbol{A}(\boldsymbol{q}_e)\boldsymbol{\omega}_{t/I}^{t}\times\boldsymbol{\omega}_{d/t}^{d}+\boldsymbol{A}(\boldsymbol{q}_e)\dot{\boldsymbol{\omega}}_{t/I}^{t}]\times\boldsymbol{p}_{d/t}^{d}-\boldsymbol{\omega}_{d/I}^{d}\times(\boldsymbol{\omega}_{d/I}^{d}\times\boldsymbol{p}_{d/t}^{d}) \\
&\quad -2\boldsymbol{\omega}_{d/I}^{d}\times\dot{\boldsymbol{p}}_{d/t}^{d}+\frac{\mu}{r_t^3}\boldsymbol{A}(\boldsymbol{q}_e)\boldsymbol{r}_t^{t}-\frac{\mu}{r_d^3}\boldsymbol{r}_d^{d}+\frac{\boldsymbol{f}_c^{d}}{m_d}+\frac{\boldsymbol{f}_d}{m_d}
\end{aligned}
\tag{C.19}
$$

实际上，对公式 $\boldsymbol{\omega}_{d/I}^{d}=\boldsymbol{\omega}_{d/t}^{d}+\boldsymbol{A}(\boldsymbol{q}_e)\boldsymbol{\omega}_{t/I}^{t}$ 进行求导可得

$$\begin{aligned}\dot{\boldsymbol{\omega}}_{d/I}^{d}&=\dot{\boldsymbol{\omega}}_{d/t}^{d}-[\boldsymbol{\omega}_{d/t}^{d}\times]\boldsymbol{A}(\boldsymbol{q}_e)\boldsymbol{\omega}_{t/I}^{t}+\boldsymbol{A}(\boldsymbol{q}_e)\dot{\boldsymbol{\omega}}_{t/I}^{t}\\&=\dot{\boldsymbol{\omega}}_{d/t}^{d}+\boldsymbol{A}(\boldsymbol{q}_e)\boldsymbol{\omega}_{t/I}^{t}\times\boldsymbol{\omega}_{d/t}^{d}+\boldsymbol{A}(\boldsymbol{q}_e)\dot{\boldsymbol{\omega}}_{t/I}^{t}\end{aligned}\tag{C.20}$$

将式（C.20）代入式（C.19），进一步整理可得

$$\ddot{\boldsymbol{p}}_{d/t}^{d}=-\frac{\mu}{r_d^3}\boldsymbol{r}_d^{d}+\frac{\mu}{r_t^3}\boldsymbol{A}(\boldsymbol{q}_e)\boldsymbol{r}_t^{t}-2\boldsymbol{\omega}_{d/I}^{d}\times\dot{\boldsymbol{p}}_{d/t}^{d}-\dot{\boldsymbol{\omega}}_{d/I}^{d}\times\boldsymbol{p}_{d/t}^{d}-\boldsymbol{\omega}_{d/I}^{d}\times(\boldsymbol{\omega}_{d/I}^{d}\times\boldsymbol{p}_{d/t}^{d})+\frac{\boldsymbol{f}_c^{d}}{m_d}+\frac{\boldsymbol{f}_d}{m_d}\tag{C.21}$$

式（C.21）即相对平动误差动力学方程。分析可知，式（C.21）实际上就是将附录 A 中式（A.5）的相对轨道动力学方程投影到从星本体系 d 系中。